John William Dawson

John William Dawson

Faith, Hope, and Science

SUSAN SHEETS-PYENSON

McGill-Queen's University Press
Montreal & Kingston • London • Buffalo

© McGill-Queen's University Press 1996
ISBN 0-7735-1368-X

Legal deposit first quarter 1996
Bibliothèque nationale du Québec

Printed in Canada on acid-free paper

This book has been published with the help of a grant from the Social Science Federation of Canada, using funds provided by the Social Sciences and Humanities Research Council of Canada. Funding has also been received from Concordia University.

McGill-Queen's University Press is grateful to the Canada Council for support of its publishing program.

Canadian Cataloguing in Publication Data

Sheets-Pyenson, Susan, 1949–
   John William Dawson: faith, hope, and science
   Includes bibliographical references and index.
   ISBN 0-7735-1368-X
   1. Dawson, J.W. (John William), Sir, 1820–1899.
   2. McGill University – Presidents – Biography.
   3. College. 4. Geologists – Canada – Biography.
   5. Scientists – Canada – Biography. 6. Educators –
   Canada – Biography. I. Title.
   LE3.M217 1855 S54 1996   378.714'28   C95-900776-8

Typeset in New Baskerville 10/12
by Chris McDonell, Hawkline Graphics.

For my parents, Ted Charles Sheets and Martha Merrill Sheets

Contents

Preface ix

1 Dawson and the Light of Knowledge 3
2 Nova Scotia Roots 15
3 So Many "Opportunities of doing good" 26
4 A Real Horse Race 38
5 "Good results in store" 53
6 "Stand by and grumble" 73
7 "None knew him but to love him" 91
8 "One of the deepest mortifications of my scientific life" 107
9 A Mission of Popularization 119
10 "A quiet middle course" 136
11 Nova Scotia Revisited 149
12 "Mighty trees from small saplings grow" 165
13 Putting Montreal on the Scientific Map 180
14 Toward International Science 190
15 No More Toil 204

Notes 213

Index 269

Preface

This book represents the culmination of more than a decade of research into the life and times of John William Dawson. The Social Sciences and Humanities Research Council of Canada encouraged this project from the beginning, by means of a series of research and strategic grants. Publication is made possible through the kindness of the Social Science Federation of Canada and Concordia University.

I am indebted to librarians and archivists in both North America and the United Kingdom for their assistance in illuminating recondite episodes in Dawson's career. I owe a special debt to the staff of the McGill Archives, particularly to Phebe Chartrand and Robert Michel, who have remained cheerful over these many years.

A number of scholars and friends have helped with various chapters of this work, as specific acknowledgement will make clear. Others have generously given their support and encouragement from the beginning. My husband, Lewis Pyenson, brought his prodigious intelligence and unerring moral convictions to bear on the undertaking. As well, Stanley Frost offered continuous inspiration, both by precept and example. I am especially grateful to my three dear children – Nicholas, Catharine, and Benjamin – who caused me to see Dawson's life in a new way. Finally, I am indebted to Wendy Dayton for her accomplished editing of the manuscript, and to Marc Speyer-Ofenberg for preparing the index.

Senneville
On a bleak November day,
the 95th anniversary of Dawson's death

The Dawson banner, showing their crest (McGill University Archives)

Daguerrotype of Dawson's mother, Mary Rankine (McCord Museum of Canadian History, Notman Photographic Archives)

Daguerrotype of Dawson's father, James (McCord Museum of Canadian History, Notman Photographic Archives)

Daguerrotype of Margaret Mercer Dawson in 1847, at the time of her marriage (McCord Museum of Canadian History, Notman Photographic Archives)

Pencil sketch of Pictou by Dawson, ca 1845 (McGill University Archives)

Daguerrotype of Dawson in 1847, at the time of his marriage (McCord Museum of Canadian History, Notman Photographic Archives)

Pictou Academy, ca 1860 (McCord Museum of Canadian History, Notman Photographic Archives)

Dawson in 1858, shortly after his arrival at McGill (McCord Museum of Canadian History, Notman Photographic Archives)

McGill campus as it looked during Dawson's early years (McCord Museum of Canadian History, Notman Photographic Archives)

Dawson family on the steps of the Arts Building, ca 1865 (McCord Museum of Canadian History, Notman Photographic Archives)

Peter Redpath Museum before completion, ca 1882 (McCord Museum of Canadian History, Notman Photographic Archives)

Anna Dawson (McCord Museum of Canadian History, Notman Photographic Archives)

William Bell Dawson (McCord Museum of Canadian History, Notman Photographic Archives)

George Mercer Dawson (McCord Museum of Canadian History, Notman Photographic Archives)

Margaret Dawson with Eva (McCord Museum of Canadian History, Notman Photographic Archives)

Rankine Dawson (McGill University Archives)

Dawson, aged 50 (in 1870) (McCord Museum of Canadian History, Notman Photographic Archives)

The Dawson family in front of the Harrington cottage at Little Métis (McCord Museum of Canadian History, Notman Photographic Archives)

Devonian plant fossil (McCord Museum of Canadian History, Notman Photographic Archives)

Illustration from one of Dawson's articles in the *Leisure Hour* (McGill University Archives)

Insignia of the Natural History Society of Montreal (McGill University, Blacker-Wood Library)

The museum of the Natural History Society of Montreal, at the corner of Cathcart and University Streets (McCord Museum of Canadian History, Notman Photographic Archives)

Reception for the American Association for the Advancement of Science in the Redpath Museum (*Canadian Illustrated News*, 1882)

Poster for the Montreal meeting of the British Association for the Advancement of Science, 1884 (McGill University, Blacker-Wood Library)

Dawson in 1895 (McCord Museum of Canadian History, Notman Photographic Archives)

Dawson with a geology group, 1895 (McCord Museum of Canadian History, Notman Photographic Archives)

John William Dawson

1 Dawson and the Light of Knowledge

John William Dawson, writing in 1878, vowed to stay in Quebec until there was no further hope for English Protestant education. To Dawson, education provided the first secure step on the path toward enlightenment, with the lamp of science illuminating the trail's forward direction, as well as its occasional twists, turns, and deadends. Dawson believed that by keeping the flame of knowledge bright – fueling it by unwavering faith and an uncompromising morality – one could conquer ignorance, eradicate prejudice, and vanquish bigotry.

Dawson possessed a missionary zeal, coupled with the conviction that knowledge and science bring real power. A clear sense that his life's purpose had been predestined from the beginning imbued all his actions. Even his character strengthened in the face of adversity, given his tendency to draw moral lessons from each apparent setback. Moreover, his resolve turned to iron when the task at hand appeared elusive and resistant to his best efforts. Thus it was that, when faced with apparently insurmountable obstacles to guaranteeing educational privileges for the English minority, Dawson decided to remain in Quebec until his death in 1899 – a period of some twenty years. Altogether he lived in Montreal for nearly half a century, virtually his entire productive adult life. Never did he become disheartened. Never was he to abandon his unceasing struggle for the scientific advancement of Montreal, Quebec, and Canada – his "own country and allegiance."[1]

Despite his work as an educator, Dawson was to make his mark as a geologist and paleontologist. He was born to parents who had endured a grueling ocean crossing from Scotland, seeking in Nova

Scotia a deliverance from the crushing poverty of their home country. At first, Providence smiled upon the newlyweds, rewarding James Dawson's tentative entry into the vibrant world of Canadian commerce. But the easy prosperity of maritime trade could easily sour, and the young Dawson family fell prey the boom years' inevitable bust. Financial disaster shattered their domestic tranquillity. Alas, this was just after the birth of Dawson – always called "William" by his family and friends – in 1820. Thus, for the first thirty years of their son's life, the Dawsons were preoccupied with repaying their debts, a responsibility that they steadfastly discharged. Perhaps the young William's intense earnestness and self-reliance were honed in this environment, for in the struggle to make ends meet in the Dawson household – where frugality was probably as important as piety – he was to shoulder a lifelong, omnipresent seriousness of purpose.[2]

Whatever young William lacked in material wealth during his formative years was overshadowed by his parents' deep affection and the rich resources of the seaside town of Pictou. Culturally speaking, Pictou could claim an academy that offered Dawson and other youths a remarkable grounding, and in a range of subjects, especially the natural sciences. Moreover, the town's environs of rich sandstone and coal formations provided fertile ground for Dawson's first scientific explorations; eventually, they would bring him into contact with the leading geological minds of the day, particularly Charles Lyell and William Logan.

The twenty-year-old Dawson's world expanded enormously in 1840, when he decided to return to his parents' homeland, where he enrolled at the University of Edinburgh. Although he remained fiercely proud of his birthplace and sensitive to any charges of hailing from the "backwoods," Canada must have seemed impoverished compared with the rich tapestry of life in Edinburgh, a city still riding the fame of its golden years. Financial difficulties called him back to Pictou shortly thereafter, but he was subsequently to return to Edinburgh to extend his studies and seek a wife. Indeed, so smitten was he with Edinburgh and its renowned university, that he was later to apply there twice for posts.

After his return to Nova Scotia in 1847, Dawson became involved in practical applications of his scientific training, in both mining and agriculture. During the early 1850s, he travelled the length and breadth of Nova Scotia as its first superintendent of education, at the same time making some of his most important paleontological discoveries. In 1855, Dawson left his native province to accept the unexpected offer of the principalship at McGill University in Montreal, Quebec.

In Montreal, Dawson became the indefatigable architect of McGill's rise to educational eminence. Dawson's Montreal years, from 1855 through 1899, provide the chief concern of this biography. Decade after decade, he worked to transform McGill University from a poorly equipped, provincial college, into a scientific institution of the first rank. He also successfully lobbied for the formation in 1882 of a national scientific organization – the Royal Society of Canada. As well, he was responsible for bringing both the American and British associations for the advancement of science meetings to Montreal, during the early 1880s. His firm guidance was felt, too, in the affairs of the Geological Survey of Canada, as well as in myriad scientific societies, especially the Natural History Society of Montreal.

MAN OF SCIENCE AND ENLIGHTENED ADMINISTRATOR

Although this study of Dawson started out as a scientific biography some years ago, it has evolved in the course of research and writing into a fuller portrait of a man of science. It seemed to me that unless Dawson's personality and character were evoked, this tale could become one of those "curiously bloodless affairs," like most of his biographical treatments to date – works about Dawson's life, but themselves without life.[3] Such an approach is particularly misleading in the case of Dawson, whose humanity and social conscience informed many of his contributions. As another historian puts it: "A decision to describe personality without science or science without personality, or philosophy, or political and social activity, is a decision that robs biography of most of its significance."[4]

Indeed, for Dawson himself, science was a comprehensive term, a multifaceted activity. To be scientific constituted a proper approach to the universe, whether this orientation was displayed in the world of daily affairs or in the workings of the mind. Dawson not only adopted the dispassionate, rational comportment of the scientist while examining a paleontological or geological specimen, but also while attempting to untangle some educational matter or delving into the intricacies of Biblical exegesis. He even reprimanded the prime minister of Canada, John A. Macdonald, for mistakenly addressing him as "Reverend" on one occasion, explaining that he was a "working geologist" (perhaps even the oldest living British North American who had selected that profession deliberately and gone abroad for scientific training); he explained that he did not like to be confounded "with the numerous Reverends and other gentlemen who have taken up geology as amateurs."[5] In Dawson's view, these latter not only did not share his rigorous methodology but tended to act on faith and from emotion. That was not Dawson's way.

In many respects, Dawson was a proper Victorian gentleman for whom "a change is as good as a rest"; as a result, his intellectual and social commitments were astonishingly large, varied, and unceasing. Richard Westfall's phrase "never at rest," used to describe the life of Isaac Newton, epitomizes Dawson's life as well,[6] as does Michael Faraday's maxim, "Work, finish, publish." Like successful scientists of all times, Dawson was exceptionally ambitious, energetic, and hardworking; in modern parlance, "driven," a "Type A personality," or a "workaholic."[7] Indeed, one British writer, a contemporary of Dawson, saw his varied pursuits as exhibiting the "natural versatility" characteristic of colonials who were called upon "to play many parts," an assessment that Dawson himself shared.[8]

Certainly Dawson's responsibilities and duties at McGill University alone were staggering, even viewed a century after his resignation as principal in 1893. He recounts in his autobiography that he regularly delivered about twenty lectures weekly during his early years in Montreal, besides superintending the university at large, which, under his direction, had embarked on an era of expansion. As well, shortly after his arrival in Montreal from Pictou, he was asked to direct the McGill Normal School and to lecture on natural science there. As a result of this unforeseen task, he estimated that he lost two months every year, periods originally promised as free time for his own research.[9] Nonetheless, he still managed to publish an average of ten scientific papers a year, quite apart from his numerous popular books and articles on educational, social, and religious issues.

Only after logging nearly thirty years at McGill did Dawson finally manage to take a long leave of absence in order to tour the Holy Lands and, on his return, to reduce the range of his teaching responsibilities.[10] At age seventy, he was still delivering fourteen lectures a week in geology and zoology.[11] Furthermore, after his retirement, ten people were required to replace him.[12] Even then, Dawson's habitual walk across the McGill campus, bag in hand, became a daily sight, for he began rearranging the natural history collections at the Peter Redpath Museum. Up until a few days before his death in 1899, he was busy editing a scientific paper.[13]

It is hardly surprising that Dawson wondered whether a position in the United States or England might mean less work and more time. He feared that "my strength is to be spent here, in building up stone by stone institutions which in their full benefits will be enjoyed only by those who may come after me."[14] The theme of his incessant toil, particularly on behalf of McGill, was to become almost a *leit-motiv* in his correspondence. He complained to his British colleague William C. Williamson, with whom he shared a special interest in the paleobotany of the coal measures, that he was "chained to the college oak," unable to escape to pursue his scientific interests.[15] Another McGill professor wondered how Dawson found time for anything else, given the "many persistent and irritating trifles" that accompany the principalship.[16] Fifteen years before retiring as principal, Dawson warned McGill's chancellor that he was considering resigning "both in the interest of my chances of life and of the special scientific work which I long if possible to complete as fully as I can."[17] The constant strain of "nothing but red tape and interruptions" prompted him to run off whenever possible to "Birkenshaw," his summer home at Little Métis, on the lower St. Lawrence River.[18] Even there, recollected his granddaughter, his "mind never rested," for he was exploring "rocks, minerals, sea creatures and plants."[19]

In his "scientific" approach to all these activities, Dawson brought into play the organizational talents of a twentieth-century administrator. In so many areas – whether apropos the affairs of the Natural History Society of Montreal, the direction of McGill University, or the proceedings of the provincial Protestant Committee on education – chaos had prevailed before Dawson's arrival. He possessed the rare gift of being able to cull the meaningful from the meaningless, oil the machinery of a creaky organization, and orient the new and revitalized institution. Even the captain of industry Peter Redpath complained that he did not have even "a tenth part" of Dawson's systematic approach to work.[20] It was Dawson's rational activism – practically unique in his day – that allowed him to accomplish so much.

Many historians, especially those attempting to portray the intellectual aspects of Dawson's scientific work, have missed this central characteristic of his existence: the unique combination of scientific imagination and entrepreneurial skill. In their concern to explore Dawson's ideas about evolution, glaciers, or *Eozoön*, they failed to see his special ability to organize and delegate responsibility – part of his comprehensive and scientific approach to the world. They also ignored his equally important commitment to facilitating the exchange of scientific ideas. These salient character traits are shown in both his creation of a national scientific organization, the Royal Society of Canada, as well as his commitment to promoting the fortunes of the British and American associations for the advancement of science.[21]

Dawson's remarkable administrative skills did not escape the notice of his contemporaries in Montreal. In their eloquent eulogies at his memorial service, his colleagues placed him in the towering presence of Charles William Eliot of Harvard and Daniel Coit Gilman of Johns Hopkins.[22] (In his own inaugural address at McGill more than forty years earlier, Dawson had expressed his hope to transform the university into a Montreal version of Harvard College.)[23] Part of their appreciation undoubtedly stemmed from his skill as a fundraiser: it is estimated that Dawson singlehandedly raised more than six million dollars in private donations for the university.[24]

Given Dawson's importance, whether viewed from the perspective of his own day or ours, it is surprising that he has never been the subject of a full-length biography. Almost an archetypical Victorian (whether for his feverish, encyclopedic intellectual activities or his socially correct comportment in all public situations), Dawson was to escape treatment in the standard multivolume "life and letters" of his time. This is particularly puzzling given the voluminous proportions of his personal and scientific correspondence. None of his scientific disciples, his long-time associates at McGill, nor his five offspring who survived infancy saw fit to describe the life of this versatile, if somewhat rigid and domineering, individual. The sole contemporary record surviving, apart from short biographical entries in dictionaries and formal obituaries, appears in Dawson's own autobiography, *Fifty Years of Work in Canada: Scientific and Educational,* published in 1901, two years after his death. According to his own (and favourite) son, George Mercer Dawson, this was "likely to be the only extended biography" of his father, at least in the near future.[25]

George proved to be correct in his appraisal of the situation. Seventy years later, Charles O'Brien attempted to capture what he saw as the essential points of Dawson's life in *Sir William Dawson: A Life in Science and Religion.* This book, though, is an intellectual portrait that emphasizes Dawson's role as a controversialist; in it, O'Brien places Dawson firmly within a whole generation of natural historians who lived to see their conception of the field usurped by younger professionals and specialists.[26] Indeed, Dawson's passionate commitment to a universe based on the doctrines of William Paley brought him into close intellectual communion with three likeminded British geologists – John Jeremiah Bigsby, Charles Lyell, and William Logan – during their sojourns in Canada. These relationships were sustained for years afterward by means of lengthy and frequent correspondence. Yet as Dawson's religiously motivated scientific fervour faded to dogmatism, he appeared increasingly anachronistic to the younger men who were beginning to populate the surveys, universities, and museums across Canada and elsewhere by the end of the century. Their view, which places Dawson on the sidelines of geological and paleontological discourse, has been perpetuated by history, to the detriment of a more objective estimation of the man and his work.[27]

9   The Light of Knowledge

THE PRIVATE AND PUBLIC DAWSON

The process and fruit of biographical writing have provided fallow territory for psychoanalytic interpretations and reflections.[28] In the words of Leon Edel, one of the finest contributors to this literature and one who underlines the psychological motivations of both biographer and subject: "A constant struggle is waged between a biographer and his subject, a struggle between the concealed self and the revealed self, the public self and the private [self]."[29] Edel further argues that the psyche and personal experiences of the author are at least as important as those of the subject, and can account for many of the characteristics of biographies: their length, their attention to an inordinate amount of seemingly trivial detail, and even the all too frequent inability of the author to finish the undertaking.

Certainly, in the course of writing this biography, particularly throughout the long process of trying to tease out the essential components of a scientific biography, my notion of the elements of greatest importance in Dawson's life did change. My own experiences, as much as my deepened knowledge of Dawson's interests and concerns, allowed me to reshape the raw data, giving it new texture and meaning. Inevitably in the course of projects of such long duration, one realizes how one's personal priorities shift. Varied aspects of the subject's life suddenly assume special importance, leaping out to grab one's attention, while other events, previously seen as central, fade into insignificance. Indeed, to the biographer, the subject's life appears to contain the ever-shifting shapes and colours of a kaleidoscope. Dawson's role as a parent and his relationship to his children, for example, only assumed significance once I became a mother and began the long process of raising three offspring to adulthood.

In my approach to Dawson's life, I have emphasized the revelations provided by his rich correspondence. During the early 1970s, the McGill Archives was able to assemble an enormous collection of Dawson's personal, administrative, and scientific papers, due largely to the generosity of his granddaughter, Lois Winslow-Spragge. I have published an index to Dawson's letters, enriched by the addition of hundreds culled from archives across North America and the United Kingdom.[30] The existence of this huge repository of correspondence allows a more complete glimpse into Dawson's private life than was once available. To use Edel's terminology, it permits the "figure under the carpet" – Dawson's unconscious motivations – to emerge.[31] In contrast, the earlier (and, I believe, one-sided) emphasis on Dawson's scientific work was derived largely from his published articles and books; that is, from the public record alone.

Another particularly revealing source of Dawson's private motivations, fears, and goals is the manuscript version of his autobiography, although it survives only in fragmentary form. While the words represent the recollections of an elderly Dawson – turning it into what son George called "an old man's book" – they provide a striking testimony to the passions and priorities of a lifetime.[32] The manuscript is especially valuable because of several significantly different passages, presumably expunged from the published version by Dawson's youngest son, Rankine. It provides, too, a remarkable illumination of the family's interpersonal relationships. Rankine's role as editor had, indeed, provoked extraordinary discord in the Dawson circle; eventually George was to refer to the matter as "this wretched biography business."[33]

One still hesitates to place Dawson on the psychiatrist's couch; as Edel suggests, we risk making utter fools of ourselves by rushing in where we ought not to tread.[34] Yet it now becomes possible to begin to deal with the realm of Dawson's motivation – hopefully sidestepping the "esoteric realm of unfathomable drives and mysterious mental closet dramas"[35] – whereas earlier attempts would have foundered as mere speculation. Dawson's letters suggest, for example, that his pronounced egotism was undercut by a massive inferiority complex. This peculiar balance may explain, in part, his contentiousness, his acute sensitivity toward title and rank, as well as his prodigious intellectual output.

One proponent of psychobiography, Miles Shore, insists that "the highest expression of the biographer's art lies in the elucidation of the nuances of motivation and relationship which form the personal myth and make it possible to see the psychological unity within which action takes on meaning." In his thoughtful essay, Shore urges the biographer not only to recognize the important role of the unconscious, but also to take account of how guiding ideals, reactions to aging, and social pressures, for example, interact with these "unconscious processes and infantile strivings." Shore adds, further, that as psychiatric knowledge grows, the biographer will benefit from that discipline's new insights into human behaviour.[36] Thomas Kohut, another advocate of psychohistory, similarly argues that the inability to employ important psychiatric techniques, such as transference, is more than compensated for by the possibility of studying "the entire life-curve" of the individual.[37] Thus, a more extensive reading of Dawson's unpublished writings than ever before (showing, as they do, how his aspirations and dreams shifted and changed over the course of half a century) can bring about a deeper, more empathetic portrait of the man himself.

As we move into the era of total dependence on telephones, electronic mail, and facsimile copies, where intellectual intercourse is fleeting and impermanent, we may well wonder about the data for future biographies. In Dawson's case, however, one is almost overwhelmed by the embarrassment of riches. Undoubtedly, his correspondence was like all his affairs, in which, according to his daughter Anna, he was "systematic and regular in his habits," never neglecting or putting off anything, no matter how apparently trivial.[38] Anna's daughter, too, remembers her grandfather as always writing: "He would sink into his swivel chair, and write and write" while her grandmother sat beside him.[39] Son George recalled that his father always believed that getting something into print and circulating it represented "a great point gained."[40] Not only did Dawson write incessantly, but a large proportion of what he wrote has been preserved through the devotion of his wife and children. One is confronted with a situation not unlike that of Virginia Woolf's as she began to write a biography of her friend Roger Fry; she pondered: "How can one make a life out of six cardboard boxes full of tailors' bills, love letters and old picture postcards?"[41]

A recent flowering of scholarship has brought the important public events and concerns of Dawson's life into sharper focus, thereby allowing us to make sense of the plethora of information. Included are histories of McGill University, of women at the same institution, and of the Geological Survey of Canada, in all of which Dawson plays a significant role.[42] Three other works explore the place of natural history and natural theology in nineteenth century Canada, incorporating the seminal contributions of Dawson in these areas.[43] As Dawson's backdrop is better understood, he himself achieves clearer definition.

DAWSON THE INSTITUTION BUILDER

Science historian Thomas Hankins contends that "historians either fight the centrifugal tendency of science to fly off from the body of history, or ... happily fly off with it secure in their conviction that science has little or nothing to do with the rest of history anyway."[44] Most, he notes, are content to concentrate their skills on reconstructing the life of the mind, although they may broaden their ideational portrait to include the scientist's philosophical predilections and religious attachments.

In Dawson's case, given the present state of knowledge, it becomes much more difficult to consider his intellectual achievements apart from the context of his myriad social concerns, particularly his institution building. Dawson himself constantly complained of how administrative work sharply curtailed his scientific creativity and productivity.[45] Nonetheless, his contributions to the geology of the Maritimes, especially Nova Scotia, and his elucidation of the fossil flora of the Devonian formation were virtually without equal during his lifetime.

In Dawson's own autobiography, a full discussion of his research and publications (largely his controversial and somewhat discredited work on *Eozoön* and glacial geology) receives treatment in only one of fifteen chapters. The omission of even a summary of his scientific work, apart from that which dealt with education, was to prove most distressing to his son George, himself an eminent geologist.[46] Younger son Rankine, however, argued that if the autobiography stinted on describing his father's scientific achievements, the neglect was intentional. In his view, Dawson meant his autobiography to be "a completion and rounding off [of] his own life's work"; that is, he wanted to be remembered not for painstaking and original contributions to narrow scientific specialities, but as a "doer and shaker" who had transformed the landscape of science and education in Canada.[47]

This present biography follows Dawson's own assessment, emphasizing his commitment to science, rationality, and the advancement of knowledge within the context of his institutional work. Individual chapters develop this approach through a series of themes, for the most part following a chronological order. As a result of the thematic approach, certain topics that were of concern to Dawson simultaneously are treated separately. For example, his direction of McGill's academic affairs and his geological research of the same years are considered in different chapters. As well, certain topics are discussed in more than one place. For example, the creation of the Peter Redpath Museum is treated both in the chapter on McGill (chapter 5) and in that on the association for the advancement of science meetings (chapter 13). In still other instances, the thematic approach means that Dawson's life is portrayed as a series of snapshot-like vignettes, somewhat insulated from the surrounding narrative. This occurs in chapter 4, which discusses Dawson's assault on the chair of natural history at the University of Edinburgh, and in Chapter 7, which examines Dawson's personal relationships with family and friends.

What Dawson saw as his period of "early growth," characterized by "preparation and active exertion," is treated in the first four chapters of the book.[48] Chapter 2 considers Dawson's youth and upbringing, and examines in detail the remarkable secondary education that he received at Pictou Academy. It also discusses the extraordinary local environment's piquing of his scientific interests and his early collaboration with the eminent geologists, Charles Lyell and William Logan. Chapter 3 continues the sketch of Dawson's Pictou years, beginning with his courtship of, and marriage to, Margaret Mercer of Edinburgh; and subsequently treating his extensive geological and mineralogical explorations of his native province, some of which were conducted during his travels through the Nova Scotia countryside as the first superintendent of education for the province. The publication of *Acadian Geology* was the effective culmination of this phase of Dawson's life: not only did it bring together his diverse intellectual labours, but it marked the year of his departure from Nova Scotia. Chapter 4 describes a brief but remarkable episode in which Dawson, working out of Pictou, entered the competition for Robert Jameson's natural history chair at the University of Edinburgh. It is astonishing not that he lost, but that he was even a leading contender. Edinburgh's loss, of course, was Montreal's gain. Chapter 5 explores how Dawson transformed McGill from a "tiny, poverty-stricken provincial school" into a well-endowed university of worldwide reputation.[49] Special attention is given there to his promotion of the teaching of the natural and physical sciences, particularly inasmuch as these subjects made the university responsive to the Canadian cultural milieu.

Most of the book deals with what Dawson saw as his middle period of "routine and uniformity," a time of "comparative stability."[50] Chapter 6 examines Dawson's educational work in Quebec beyond the boundaries of McGill, especially as it reflects his views on the consequences of Canadian Confederation. The chapter also discusses one of his personal responses to Confederation; namely, a second unsuccessful attempt to obtain a position at the University of Edinburgh, this time the principalship. Shortly thereafter, Dawson was called to a chair at Princeton University but declined, saying that his children were well launched on productive careers in Canada. In chapter 7, Dawson's relationship to his family of five children is examined, as well as his associations with several lifelong friends.

Aspects of Dawson's intellectual activity form the substance of the succeeding four chapters. Chapter 8 describes the disappointing response to his work on the fossil flora of the Devonian formation, epitomized by the Royal Society of London's refusal to publish his Bakerian lecture. This event assumes special significance, inasmuch as it changed the direction of his scientific writings. Chapter 9 considers his views on religion and science, particularly as shown in the publication of *Archaia;* his work as a popularizer; and his uncompromising opposition to Darwinian evolution. Chapter 10 explores two other areas in which his scientific work plunged him into controversy: his writings on glacial geology and on *Eozoön*. In contrast, chapter 11 treats the relatively safer haven of Carboniferous geology and paleontology, specifically fossil reptiles, mining, and subsequent editions of *Acadian Geology*.

The next three chapters deal with Dawson's work as an institution builder. His revitalization of the sagging fortunes of the Natural History Society of Montreal, and its transformation into a dynamic amateur scientific organization, form the substance of chapter 12. Chapter 13 examines Dawson's role during the early 1880s in bringing the meetings of the American and British associations for the advancement of science to Montreal, events that also enhanced the reputation of the Natural History Society. Chapter 14 outlines the shifting of Dawson's organizational talents to the national level, as reflected in his creation of the Royal Society of Canada and his subsequent attempts to initiate international scientific cooperation.

Because of this range of interests and Dawson's virtually incessant activity right up to a few days before his death, it is difficult to identify the beginnings of the later period of his life, a period that he compared to "a ripening of fruit, and a sifting ... of the grain from the chaff." This was an era, he said, "of culmination in some respects, and decadence in others."[51] In retrospect, one might point to his stubborn adherence to creationism or to his rigid stance concerning the nature of *Eozoön*, as indicators of the onset of mental, if not physical, old age. His falling out with the leadership of the International Geological Union likewise appears unlikely for someone like Dawson, so practised in the art of scientific diplomacy. After a severe bout of pneumonia in 1892, however, he became infirm physically and, in 1893, stepped down as principal of McGill. At this point, he began to spend his winters in the southern United States. Chapter 15 considers all these matters, offering a few remarks on the lessons to be drawn from Dawson's struggle for the scientific advancement of Canada.

2 Nova Scotia Roots[1]

For John William Dawson, only two places could be called "home" during his lifetime of seventy-nine years: Pictou, Nova Scotia, where he lived until age thirty-five; and Montreal, Quebec, where he moved in 1855 to become principal of McGill University. Little did he dream when he left Nova Scotia that he would never again live in the picturesque coastal town where he was born. Yet the legacy of Pictou was to remain a strong, formative influence: beneath Dawson's half-century of varied and numerous accomplishments in Montreal lay the deliberate design of a purposive and tenacious personality. The regularity with which the same traits were exhibited may have made him seem consistent and dependable to his friends, but to his foes he appeared stubborn and unbending.

At first glance, it appears somewhat surprising that the values inculcated in that small, culturally homogeneous maritime village should have stood Dawson in good stead in such a radically different urban milieu. But in relatively polyglot Montreal, the Roman Catholic faith of the majority of its inhabitants would prove to be more troubling to Dawson, the staunch Presbyterian, than would the actual foreignness of the city's French or Irish. During the early nineteenth century, furthermore, Pictou and Montreal were not the sharply contrasting environments that they are today. Pictou then was a major port city with a "capacious and well-sheltered harbour," where three rivers united and joined the waters of the Northumberland Strait.[2] During the first two decades of the nineteenth century, its exports, principally timber, averaged £100,000 a year, sometimes totalling more than £300,000.³ As for the island city of Montreal, it was at that time provincial and isolated, lacking even the bridges that would eventually connect it by rail to the outside world.⁴

If one believes the sketch of the Pictou temperament depicted in a local newspaper, Dawson's personality traits owe much to the environment in which he was born and raised. According to the article, the "Pictovian type" belonged to the Calvinist genus, with its stress on the dogmatic and ascetic, rather than the genial and urbane. Yet the species – being of Scottish descent – also meant the predominance of two characteristics: ambition and an attachment to education. Add to that the imagination and adventure of the "Celtic" or Highlander variety that dominated the migration to Pictou, as opposed to the Calvinism of the Lowlanders, with their dour and sombre traits (allegedly shared with the Dutch and New Englanders). In fact, then, given his boundless energy, practicality, sturdy individualism, personal integrity, and capacity for organization, Dawson perfectly embodied the "Pictovian type."⁵ These qualities were further enhanced by a deeply ingrained sense of divine calling and an "unflagging sense of duty," said to be typical of Nova Scotia Protestants.⁶

According to these assessments, Dawson's temperament stemmed from genetic predisposition, enhanced by the religious convictions of family and friends. Yet "nurture" was to play as important a role as "nature" here, for one sees the direct impress of the Pictou environment on many aspects of the young Dawson's character. Not only was Dawson's personality shaped by the natural world around him, but his capacities and interests were profoundly influenced by the cultural resources of the town. An examination of Dawson's youthful years in Pictou will reveal the range of unique circumstances that so powerfully shaped his future career.

THE DAWSON HERITAGE

In 1811, after five weeks of travel in steerage,⁷ Dawson's father, James, arrived in Pictou from Banffshire in the North of Scotland. The town of Pictou had just emerged from the status of backwoods outpost, settled by immigrants from Scotland, England, and the United States. As the younger Dawson was later to describe the cultural mix, there were "persons of good education and of reputable antecedents, mingled with all sorts of waifs and strays."⁸ By then, the population numbered more than 5000, and about twenty buildings served as the village core. Pictou's most valuable natural resource was the white pine, so quickly squandered by early settlers, that covered the hills of the surrounding countryside.⁹

From the beginning, James Dawson tested the commercial waters in Pictou by dabbling in "Indian Porcupine Quill Manufacture" and the fur trade. By 1813, having left his apprenticeship as a saddler, he embarked on more ambitious entrepreneurial undertakings. His subsequent career as a maritime merchant followed a then-typical Pictou pattern. Underpinning it was the abundant local timber, particularly the desirable "squared pine," which could be used in local shipbuilding or exported to Britain and elsewhere. Ships to carry out the export trade were built in Pictou, as were small vessels to fish the teeming coastal waters. Both the fish and surplus timber were carried to the West Indies, from whence cargoes of native produce returned to Nova Scotia.[10]

At first, James was successful enough to send for his brother Robert, for he needed a partner in these mercantile concerns. But by the mid-1820s James experienced a series of business reversals that would lead him to compare himself to Job. Although James's experiences were especially unlucky and severe, they also reflected, at an individual level, the crippling financial crisis of 1823-24 that afflicted everyone in the timber trade and shipbuilding at the time, whether in Pictou or the home country.[11]

In the meantime, James had married Mary Rankine, about whose early life little is known. Mary, who came from Stirlingshire but whose parents had died when she was young, was raised by "distant relatives in Edinburgh," perhaps the Boyds of the well-known Edinburgh bookselling firm, Oliver and Boyd.[12] Perhaps, too, this family connection was part of the reason why James made his fortuitous decision to turn from the sea to the printing, bookselling, and stationery trade. In any case, he opened the first such business in Nova Scotia outside Halifax, which allowed him eventually to repay almost all his earlier debts. His decision to concentrate on stocking much-needed books, in contrast to his Haligonian counterparts who preferred to feature more frivolous items in the stationery line, led to a modest success.[13] Not only did James become proprietor of the local newspaper, he also acted as principal inspector of books (and other literature) for eastern Nova Scotia.[14]

But James Dawson's commitment to education extended far beyond simply supplying tools in the form of books to educators; it was enhanced by his deep religious convictions. Shortly after his arrival in Nova Scotia, he began the first Sunday School in Pictou County, probably the only schooling available to many children in the area. Undoubtedly, the importance he attached to circulating useful knowledge and promoting popular education was to exert a profound influence on the elder of his two sons, John William, born in 1820.

William, as his family called him, grew up in a household that valued education but where money was scarce (it took his father a quarter of a century to pay back his creditors). At an early age he learned the technicalities of book publishing: he assisted his father in typesetting and bookbinding, even writing paragraphs and short articles. Yet his relationship with his father was certainly not all work. He fondly recalled morning and evening walks, where James, Sr, recited "scraps of poetry, anecdotes, and wise advice."[15]

William later remembered as the single most important event of his youth, the death in 1837 of his only sibling James, four years younger, who contracted scarlet fever as an adolescent. His death cast a long shadow over the household. Dawson recalled that it "made the world seem black for a time,"[16] and his mother suffered severely, never fully recovering her mental and physical health. Effectively an only child, Dawson found that his parents "clung to me as their last earthly hope."[17]

The filial responsibilities that this position entailed weighed heavily on the conscience of young William, who already tended to be "shy and solitary."[18] While his religious fervour deepened, his future career prospects were being called into question. When James was alive, he had been the natural choice as his father's future business partner; now, however, it was no longer obvious that William could choose a professional pursuit such as science or the ministry, for that would require long absences from the Dawson homestead.

THE PICTOU ACADEMY

Despite the modest circumstances in which William was raised, his family did manage to send him to the local grammar school, run by the widow Cameron "on the plan of the parish schools of Scotland."[19] From there, Dawson moved to the lower levels of the Pictou Academy, which proved to be an especially fortunate choice. A unique and remarkable educational institution, it was without equal elsewhere in the Maritime provinces and throughout Canada.

Pictou Academy had been established in 1817 by the secessionist Presbyterian minister Thomas McCulloch, an unexpected Pictou settler of 1803. Once established in Nova Scotia, McCulloch intended to create a school both to train dissenting ministers and to provide a liberal education for youths of all religious persuasions. The two aims could be accommodated in the same institution, he decided, because the Presbyterian church saw a grounding in literary and scientific matters as essential for its preachers. However, the complicated religious and political arrangements prevailing in early nineteenth century Nova Scotia thwarted McCulloch's plans. Pictou Academy was never to win the permanent government support that McCulloch sought for over twenty years, nor was it ever empowered to grant college degrees.[20]

But for a young man such as Dawson, Pictou Academy provided a superb education. What government parsimony had denied to McCulloch and the early trustees, the townspeople of Pictou (as well as their co-religionists in Scotland) gave privately, even when hard financial times made such generosity nearly sacrificial. The academy was able to offer a liberal, collegiate education to Dawson and hundreds of young men like him. It particularly attracted those youths who either could not afford the high cost of King's College in Windsor, Nova Scotia, or who could not abide its High Church elitism.[21]

Fifteen-year old William Dawson was registered as a student at the Pictou Academy for the academic year, 1834-35.[22] He worked his way up from the grammar school, finally donning the scarlet gown of the college and graduating in 1839.[23] The curriculum stressed a basic grounding in the "three Rs" and the classics. When resources permitted, courses were also offered in geography, French, logic, moral philosophy, bookkeeping, and navigation.[24] Even elementary mathematics at Pictou Academy could be quite comprehensive; in one year it included "algebra, Euclid, conic sections, plane and spherical trigonometry," along with practical applications.[25]

During the 1830s, however, limited financial resources seem to have drastically curtailed the curriculum at Pictou Academy. Not a single student was permitted to study natural philosophy in a formal fashion during the 1832-33 academic year; and by the end of the 1834-35 academic year, advanced students were sitting for final examinations only in moral philosophy.[26] Thomas McCulloch himself examined the upper levels in moral and natural philosophy the following year, at which point it was noted, "the evidences of their improvement were satisfactory."[27] Unstable finances caused student enrolments to fluctuate wildly during this decade (from a low of sixteen to a high of sixty-one), as the trustees arbitrarily increased student fees in a desperate attempt to meet expenses. In an era in which McCulloch's own son Michael was to resign over nonpayment of his salary, it is hardly surprising that the teaching of science was viewed as an expendable educational luxury.[28] The fallacy of this false economy was later revealed, however, when the Presbyterian Synod used the academy's failure to teach logic, natural philosophy, and moral philosophy during this period as justification for establishing its own divinity school, in direct competition.[29]

Nevertheless, the several islands of tranquillity at Pictou Academy enabled Dawson and other youths like him to learn the basic scientific tools of the trade, tools on a par with those available in Britain or on the continent. Most important was the school library, which, while well stocked in the works of classical authors, contained a remarkable assortment of treatises in natural philosophy and natural history.[30] Of particular import were the textbooks of those physical scientists associated with the Scottish Enlightenment of the late eighteenth and early nineteenth centuries.[31]

The library held, for example, two *Systems of Chemistry*, written by Edinburgh professors and archrivals, Thomas Thomson (1802) and John Murray (1806).[32] It also carried the works of two colleagues in natural philosophy at Edinburgh, John Robison and John Leslie.[33] The chemistry textbook *Elements of Experimental Chemistry*, written by Scottish-trained chemist William Henry in 1810, and called "the most popular and successful chemical text in English for more than thirty years," was available,[34] as were treatises from two other Scottish-born astronomers who became Oxford professors: David Gregory's *Elements of Physical and Geometrical Astronomy* (1715) and John Keill's *Introduction to Natural Philosophy* (1720).[35] In the library were certain "classic" popularizations of science, such as the Scottish-born itinerant lecturer James Ferguson's *Young Gentleman's and Lady's Astronomy* (1768), still used in British grammar schools as late as the 1840s.[36] The Scottish flavour of the library's scientific holdings was so pronounced that one wonders whether Oliver Goldsmith's stint as a medical student in Edinburgh persuaded the academy to stock his scientifically absurd *Animated Nature*.

Although the only French text from the realm of physical sciences at Pictou Academy was Legendre's *Eléments de géométrie*, natural history books there were dominated by French-speaking authors of the late eighteenth century. The academy possessed several of the forty-four volumes of Buffon's *Histoire naturelle générale*. It also held Cuvier's *Discours sur les révolutions de la surface du globe*, probably in its English translation by R. Kerr, with notes and comments by the Edinburgh professor of natural history, Robert Jameson. The only other work on geology was Swiss *savant* Jean André Deluc's textbook, translated into English as *An Elementary Treatise on Geology* in 1809. One wonders whether the future creationist Dawson read the work of this author, who claimed to reconcile the Book of Genesis with geology by demonstrating "the conformity of geological monuments with the sublime account of that series of operations which took place during the *Six days*, or periods of time, recorded by the inspired penman."[37]

Other works on natural history included James Lee's *Introduction to Botany* of 1760, which became a standard work, even though it merely translated Linnaeus's *Philosophia Botanica* into English.[38] Christian Konrad Sprengel's *Das entdeckte Geheimniss der Natur im Bau und in der*

Befruchtung der Blumen (1793), listed in the library inventory as his *Philosophy of Plants*, is a surprisingly technical and scientifically important inclusion, later cited by Darwin as a "wonderful book."[39]

More strictly belonging to the realm of natural theology was the collection of various treatises by William Paley, numbering some fifteen copies in the Pictou Academy library inventory. Paley enjoyed considerable popularity in Britain at that time, his *Natural Theology* of 1802 having already been reprinted twenty times by 1820. To Darwin, a budding young naturalist attending Cambridge University during the early 1830s, Paley's significance transcended his deistic explanations. Indeed, Darwin was to write of Paley's *Evidences of Christianity*:

The logic of this book, and, as I may add, of his *Natural Theology*, gave me as much delight as did Euclid. The careful study of these works, without attempting to learn any part by rote, was the only part of the academical course which, as I then felt and as I still believe, was of the least use to me in the education of my mind. I did not at that time trouble myself about Paley's premises; and taking these on trust, I was charmed and convinced by the long line of argumentation.[40]

One suspects that McCulloch found in Paley the perfect partnership of logic and moral philosophy, especially appropriate for training the minds of young men such as Dawson.[41]

Although Dawson may not have had formal instruction at Pictou Academy in either the natural or physical sciences, the library offered him much for his own quiet perusal, whether a full complement of basic philosophical texts or related works by scientistic authors such as Paley. Yet another remarkable resource housed at the academy was the extensive collection of scientific apparatus. Some of the items belonged to McCulloch himself; perhaps he used such items as the magic lattern, the model of the steam engine, and the brass windmill to illustrate his own popular scientific lectures.[42] Keen students, however, may have been allowed to conduct chemical experiments with the aid of bottled chemicals, a mortar and pestle, hydrometers, brass scales and weights, as well as an assortment of rods, funnels, retorts, crucibles, bottles, and tubes.[43] A Leyden jar, "electrifying machine," magnetic and pneumatic apparatus, barometer, several air pumps, and a prism, as well as pulleys, balls, and mirrors, served the aspiring physicists. The academy owned, in addition, a microscope, two telescopes, two orreries (though incomplete), two sextants, and an astrolabe. Such resources placed Pictou Academy on a par with the most advanced German *gymnasia* of the day, and even with some of the German universities.[44]

Thus, the young Dawson may well have had the use of the finest scientific library and collection of instruments anywhere in Canada.[45] In addition, McCulloch had established his own personal natural history museum at Pictou Academy, with his collection of local specimens and some typical British fossils.[46] The acclaimed wildlife artist John James Audubon, who saw McCulloch's collections on his way back from Labrador, estimated their worth at around £1000.[47] Dawson probably spent many hours at the museum; one of McCulloch's sons even trained him in the art of preparing natural history specimens for exhibition.[48]

Besides providing a good basic education and extracurricular opportunities for scientific study, Pictou Academy taught Dawson lessons and gave him models to which he could later refer with profit. In any event, his student years overlapped with McCulloch's last years, for McCulloch was to leave Pictou in 1838 to become principal of Dalhousie College in Halifax. Perhaps Dawson remembered McCulloch's brilliant polymathic abilities (it is said that he taught not only logic, moral and natural philosophy, political economy and chemistry, but also Greek and Hebrew) when he himself was called upon to teach virtually every subject in the curriculum during his early years at McGill.[49] Decades later, when establishing a library, a natural history museum, and proper scientific instruction in Montreal, Dawson may again have been inspired by McCulloch's example.

The complex political and religious circumstances surrounding the establishment of Pictou Academy may have provided the young Dawson with another, albeit negative, example. The turbulent early history of the academy, particularly the rancour displayed by its enemies, may well have caused Dawson to later insist that McGill be completely nonsectarian (that being the tradition of the Scottish universities).[50] As well, it may have inspired him to avoid the hand of politicians altogether, however welcome the public funds might be. After all, they had only added fuel to the flame of religious strife in early nineteenth-century Nova Scotia, with serious consequences for the cause of education.

THE GEOLOGICAL ENTICEMENTS OF THE NOVA SCOTIA LANDSCAPE

McCulloch's development of the Pictou Academy created a unique intellectual environment in early nineteenth century Pictou.[51] Living in the dynamic coastal town and attending the academy undoubtedly helped to propel Dawson toward the world of science and the life of the mind. Indeed, his parents had already prodded him in a scholastic direction, encouraging him to explore the natural world at his doorstep. As Dawson points out in his autobiography, he was especially attracted to the shale and sandstone beds containing Carboniferous fossil plants, which lay exposed in "quarries, road-cuttings, and coast and river cliffs" about the town.[52] His own geological and paleontological collections – which won the praise of McCulloch but the ridicule of his childhood friends – were further expanded by searching imported cargoes of limestone (destined for the limekilns) for marine fossils, including shells and crinoids.[53] At age sixteen he even delivered a paper on "The Structure and History of the Earth" to the Pictou Literary and Scientific Society, of which McCulloch had been a founding member.

During the 1830s, the teenaged Dawson seized the opportunity to travel to the coastal cliffs of the South Joggins, bordering Nova Scotia's Cumberland Bay (an arm of the Bay of Fundy). He expressed his amazement at "the grand succession of stratified beds exposed as plainly as in a pictured section, and ... the beds of coal, with all their accompaniments, exposed in the cliffs and along the beach, the erect trees represented by sandstone casts, and the numerous fossil plants displayed in the beds." This first visit to "the Joggins" – its peculiar name derived from the Micmac Indian language – heightened Dawson's already keen interest in geology.[54] Other geological excursions from Pictou included trips to Cape Blomidon and the Minas Basin at Truro, another product of the Fundy tides. Dawson sharpened his eye and trained his mind by consulting the standard geological treatises of the day: Charles Lyell's *Elements of Geology* and *Principles of Geology*, Henry de La Beche's *Manual of Geology*, and John Phillips's *Elementary Geology*.[55]

Young William realized, however, that the resources of Pictou could not quench his thirst for philosophical enlightenment. In 1840, at age twenty, he travelled to the University of Edinburgh, one of the few universities in the English-speaking world that offered a systematic natural history curriculum. There he learned geology, physical geography, and mineralogy from Robert Jameson, botany from John Hutton Balfour, and chemistry from William Gregory. He delivered a paper on an unusual field mouse found near Pictou to the city's Wernerian Society, an amateur scientific society dedicated to geology and natural history. Unfortunately, his family's limited financial resources were to necessitate his return to Pictou and the family bookselling business at the end of the academic year.

From the standpoint of scientific advancement, however, his return could not have been better timed. In the summer of 1841, Dawson (by then Pictou's leading amateur naturalist) met the distinguished British geologist William Logan. Logan was returning to England after a visit to the coalfields of Pennsylvania. At the time, being about to assume the directorship of the Geological Survey of Canada,

Logan was particularly interested in the important regional formations of British North America. He wrote of the Joggins: "I have never before seen such a magnificent section as is there displayed. The rocks along the coast are laid bare for thirty miles, and every stratum can be touched and examined in nearly the whole distance."[56] He also acclaimed the geological skills of Dawson, pointing out that the complicated fault lines of the Pictou coalfields made it seem "a sort of gigantic Chinese puzzle," a test for even the most capable, mature scientist. It remains a mystery why Logan did not tap the promising talents of young William when looking for personnel to staff the fledgling Canadian survey.[57] (Possibly there was personal friction: ever after, Charles Lyell was to assume the role as liaison between Logan and Dawson, even when both Logan [as Geological Survey of Canada director] and Dawson resided in Montreal).

While Dawson was to maintain a cordial relationship with Logan, he became the lifelong protégé, confidant, and disciple of Lyell, whom he also seems to have met in the summer of 1841, when Lyell alighted for six hours in Halifax *en route* to the United States. Apparently Lyell was entranced by the geological prospects of the region, because he vowed to return to Nova Scotia before departing for England. Apparently, the geology of the province – particularly the erect petrified trees – enticed him to spend a month there the following summer. On that occasion, Dawson guided Lyell through the coalfields near Pictou, the cliffs of the Shubenacadie River, the deposits at the Bay of Fundy, and the shores of the Minas Basin.[58] Afterward, they exchanged data, opinions, and advice by mail, all of which resulted in a steady stream of publications by Dawson on the geology and paleontology of Nova Scotia, New Brunswick, and Prince Edward Island.

Through Lyell's patronage,[59] Dawson began to communicate his findings to the Geological Society of London and to other scientific institutions, such as the British Association for the Advancement of Science.[60] Lyell made sure that Dawson's papers were placed on specific agendas at the times when they would receive the most attention; he even delivered them himself whenever possible. (One may discern Lyell's influence, as well, in the leading position assigned to Dawson's papers in the Society's *Quarterly Journal.*[61]) Lyell found letters to be a poor substitute for conversation, but he reported as best he could the discussions evoked by Dawson's work. For example, he described how he had defended Dawson against one critic who had remarked that there always seemed to be a break in the geological sections supplied to illustrate Dawson's papers, and invariably at the most critical point.[62] During Lyell's absences, Dawson was to report that he had "no geological friend in London."[63]

The relationship between Lyell and Dawson functioned to their mutual benefit. Dawson – who had earlier complained to a correspondent that he lacked the necessary books and collections to determine his specimens accurately – found in Lyell someone who could place his fossils in the most expert hands. Lyell recruited William Lonsdale, for example, to assist with corals; George B. Sowerby, Sr, with shells; and his own brother-in-law, Charles Bunbury, with plant fragments. In turn, Lyell, inundated with nearly forty boxes of specimens after ten months of fieldwork in North America, appreciated Dawson's communication of any "facts, generalizations, or speculations" about their findings.[64] Dawson could be entrusted to carefully pack and ship whatever specimens Lyell might still need, unlike fellow geologist and mining entrepreneur Richard Brown, whose box of fossils arrived in such damaged condition that Lyell could scarcely identify the contents.[65]

Lyell was firmly persuaded of the importance of the field of investigation that stretched before Dawson in Nova Scotia, and of Dawson's ability to make significant discoveries there. He exhorted Dawson to hunt for the footprints or remains of a coal reptile at Horton Bluff, and to find better specimens of the Joggins' vertical trees so as to help settle certain disputes of European paleontologists. Even a careful study of which Carboniferous fossils prevailed in different regions of Nova Scotia would be invaluable, he said, for comparison with those of Europe.[66]

Inspired by his contacts with high scientific culture in Edinburgh and stimulated by these collaborations with Lyell, Dawson turned to his "good friends geology and the other 'ologies'" during those few moments he could spare from the business of selling books. Perhaps because he often contrived "to mingle a bit of business with his geological excursions," he seldom hesitated to leave Pictou for a week at a time to ramble through the Nova Scotia countryside. Moreover, slow economic times in Pictou during the early and mid-1840s – at times Dawson expressed reluctance to fully stock the bookshelves at the family store – meant more leisure for reading and thinking.[67]

Certainly evenings, free days, and the more tranquil summer months found Dawson occupied with natural history pursuits. He tended to scorn the company of his peers in Pictou. Indicative of his contempt was the time when, lecturing to the Pictou Literary Society and Mechanics' Institute on geology and mineralogy, he dismissed the audience's interest as "of little consequence," particularly as compared to the benefits he himself had reaped from the process of preparing the lectures.[68] But there was one individual, a young woman he had met in Edinburgh, who began to inspire William both to forsake his solitary ways and to focus his quest for self-betterment.

3 So Many "Opportunities of doing good"[1]

Nearly a decade his junior, Margaret Ann Young Mercer (a distant cousin and daughter of a lace merchant) was but a child when she first met William in Edinburgh, apparently in 1841. Margaret's parents were delighted to have their fourth, and by much the youngest, daughter kept "occupied" for afternoons on end, during this and Dawson's subsequent visit to Edinburgh. They assumed that she would be educated and enlightened by William's conversation. But when he turned out to be a suitor intent on marriage, his relationship with Margaret's parents soured.[2]

THE COURTSHIP AND MARRIAGE OF MARGARET MERCER

Perhaps it was fortunate for the future of their relationship that Dawson was forced to conduct their courtship at a distance, by letter. The correspondence between William and his future wife is striking in its revelations of Dawson's pastimes and passions during the 1840s, when he was still in his twenties.[3] (Even the usually reticent William was struck by his own spontaneity in the letters he sent to Margaret.) He wrote to her "like a Catholic to his confessor, revealing all my thoughts and plans," even the mundane details of his daily routine.[4] He described to Margaret his filial duties: in addition to attending his father's bookshop from morning until night, both selling books and keeping his father's accounts, he looked after his infirm mother, as

27 "Opportunities of doing good"

well as several tenants who rented houses from his father, thereby giving the elder Dawson time to devote to his farm. After "shop-shutting" time, William turned to his scientific pursuits and, on occasion, studied Hebrew and even singing. He viewed dancing, however, as a questionable activity and said he looked forward to the time when waltzing went out of fashion.[5]

Dawson was no mere colonial suitor going hat-in-hand to the fount of home-country culture. His intense pride in things Canadian emerged even at this early date, apparent in the objects with which he showered his rather complacent "intended" and her family. He dispatched Indian moccasins and baskets (worked with porcupine quills and moose hair); books and magazines published in North America, and produce from the Dawson farm (including Indian corn which, in its dried form, denied Margaret the pleasure of tasting "that American luxury," as well as black currents preserved with maple sugar). He even managed to transport a rocking chair and parlour display of stuffed birds, accompanied by instructions for their proper exhibition on a stand made of Nova Scotia maple. He admitted – albeit tongue-in-cheek – that given his predilection for products of his own country, he was surprised that he preferred Margaret to a Nova Scotia lass.

Dawson scorned the British who spoke of the American colonies as nothing but "backwoods."[6] He maintained that they underestimated a fine place like Nova Scotia, whose rural population was exceptionally upstanding – not overworked and underfed like Britain's labouring classes, but well dressed and constantly seeking self-improvement.[7] Dawson argued that while the town of Pictou might claim no inhabitants of great rank or wealth on a par with those of Edinburgh, its respectable citizenry could still be characterized by their "intelligence and a desire for knowledge."[8] Perhaps the only drawback was that few in the new country had the leisure or taste to attend to its natural beauties.[9]

An intense seriousness of purpose manifested itself in the twentyish Dawson. Although he toyed with the notion of entering the ministry, he felt that he could forsake neither his father's business nor his own geological pursuits. For example, he declined opportunities to participate in geological explorations when they conflicted with responsibilities to his parents or his father's business. During one New Year's Eve spent writing an article, he reflected soberly amid the merriment surrounding him (sounding like a typical Scots Calvinist) that "it would better become most to be sad, because the past [year] might have been better spent, and the coming [year] involves much responsibility, and more uncertainty."[10] Yet even he was puzzled as to why he so preferred devoting his leisure to useful pursuits rather than to idle pleasure like most young men his age. Perhaps his strong sense of duty to parents, country, and God overwhelmed other gratifications.

In the spring of 1847 (at the completion of his second academic session at the University of Edinburgh), Dawson finally persuaded the seventeen-year old Margaret to give up home and family, marry him, and transplant herself to Nova Scotia. Although the bride and groom each had one attendant, no other friends and relatives but Margaret's disapproving parents attended the April wedding. Margaret declined the customary white gown, preferring a "pale grey dove" colour as being more suitable for the "wild land she was going to."[11]

The newlyweds sailed from Glasgow to Halifax, where they touched ground again on a beautiful summer's day. To Margaret, the buildings seemed to be made of cardboard, so different did they appear from solid Edinburgh stone.[12] After a long trip overland to Pictou, they settled in the home of William's parents. While Margaret's relationship with her father-in-law and his relatives was loving and unqualified from the beginning, that with her mother-in-law appears to have been cordial but never warm. Indeed, many years earlier, Margaret's own mother had met and disliked William's mother, who by now had grown fiercely protective of her only surviving son.[13]

Although William had promised Margaret that she would not be forced to "vegetate" in her new role, his career ambitions often took him from her side. Despite previous assurances to Margaret and her family of his steadfast devotion to the bookselling business, the late 1840s and early 1850s saw Dawson gradually enlarging his sphere of interests. Clearly, he was not content to remain a bookseller in Pictou, however comfortable an existence it would secure for his young family.

He tried his hand at other possible occupations which, at the time, offered only temporary employment. Preaching and teaching (as well as geologizing and mining) all tantalized young William, as much for their potential to transport him to new worlds, whether physical or intellectual, as for their capacity to provide useful activities in their own right. Opportunities to proselytize were presented during his mining explorations and educational tours through the Nova Scotia countryside. For instance, wherever he found a Protestant church, he attempted to assemble the congregation. If this plan failed, he managed at least to attend a church service. He also distributed religious tracts to backwoods inhabitants, especially to Roman Catholics who afforded prime targets for conversion. But teaching and advancing the cause of education seemed to capture his fancy above all else. While these tasks permitted him to indulge his passion for exalting nature and praising God, they also enabled him to engage the hearts and minds of those around him. Educational pursuits offered the ideal combination, permitting him to satisfy the material needs of his family while at the same time fulfilling his sense of divine calling.

"A GOOD DEAL OF ZEAL FOR EDUCATION"[14]

In 1848, Dawson agreed to deliver a series of lectures on natural history at the Pictou Academy. After paying the costs of a janitor and lighting, William earned nearly £16 from the fees collected from the fifty-three students who registered for the course.[15] Two years later, he began to teach on a casual basis in Halifax. He spoke to members of the Mechanics' Institute about the composition of soils, and agreed to help analyse minerals in their museum.

A more impressive invitation came from Dalhousie College, where he gave forty lectures on botany, zoology, mineralogy, and geology in 1850. Although he complained of the round of balls, parties, dinners, and meetings that kept the ninety-odd students at the college from attending his lectures on a regular basis, any empty seats were occupied by interested townspeople. He illustrated the lectures with his own drawings and natural history specimens, sent from home by the solicitous Margaret (her job was to send the required stuffed birds, microscopic slides, and slices of fossilized wood from his collections).[16] A grateful class penned glowing testimonials to his lectures and presented his new bride with a china teaset, a welcome gift when attractive and relatively inexpensive worldly goods could be found only in Halifax.[17]

Dawson seems to have expected little financial return from these engagements, particularly from the Halifax lectures, despite the considerable personal toll that his absence exacted on the young household. He believed, instead, that in Halifax he could forge important contacts, especially with members of the conservative society of official Halifax.[18] Indeed, this hope was realized almost immediately: the Provincial Secretary of Nova Scotia, newspaper editor Joseph Howe (who also sat on the Board of Dalhousie College), asked Dawson to become the province's first superintendent of education. Initially, Dawson declined the offer, fearing that he was too young and inexperienced, and that his special interest in natural sciences had made his knowledge of mathematics and classics obsolete.[19] He wrote to Margaret that he did not wish to be constantly absent from home, especially for only a temporary position, and that, although attractive, the income would still mean much discomfort for them.[20] Even his father confided to Howe that William's reluctance might make him ill and inefficient.[21] His father need not have worried about inefficiency:

Dawson, once having changed his mind, threw himself into his new job with the enthusiasm he exhibited toward everything he undertook.[22]

By May, Dawson outlined to Howe the conditions under which he would accept the job; the major one being that Pictou, not Halifax, would be his headquarters.[23] Dawson disliked that "wooden city" of 25,000, whose parliament resembled Britain's "seen through the wrong end of a telescope." He deplored its "petty aristocracy of government officials," who possessed little scientific or literary taste, but plenty of foppery.[24] Undoubtedly, he also did not want his wife, young son George, and baby Anna to leave the protection of the elder Dawson household (their first child, James Cosmo – a "dear little boy" – had recently died in infancy).[25]

Using Pictou as a base, William travelled on foot, by boat, and on horseback to visit schools and to interview teachers and parents all over the province. The conditions he endured were worse than primitive: it was not unusual for him to travel on roads made of nothing but large stones, and to use dilapidated ferry boats to cross water so rough that his horse was terrified. Often, he covered long distances in the morning, held teachers' meetings in the afternoon, and delivered lectures to the local populace in the evening. His observations and sketches of the towns he visited, along with his descriptions and candid evaluations of the customs, religious beliefs, and practices of their inhabitants, make his letters a superb chronicle of life in mid-nineteenth century Nova Scotia. He described the children of French fishermen of Cape Breton, for example, as being kept in ignorance by both priests and traders, who sought to fortify their control by excluding the schoolmaster.[26] He wrote of log houses that functioned as schoolrooms in the midst of the bush (where newspapers replaced books as reading material), and of other schools temporarily closed so that hungry children might forage for berries.[27]

After two years on the job, Dawson had visited more than 500 schools, given 113 lectures, and held 56 public meetings. In addition, he had organized eight teachers' institutes (attended by more than two hundred teachers), authored several educational works, and corresponded about pedagogical topics with a large number of individuals. The benefits of these labours, Dawson wrote, were considerable: improved teaching methods in most schools, a better supply of books and apparatus, revamped school houses, increased attendance, organization of a teachers' association, rationalized administration of the schools, establishment of school libraries, and the collection of accurate, province-wide educational statistics. Also, the public at large seemed more concerned about these issues now that Dawson had raised their consciousness.[28]

Dawson's intense commitment to the task at hand had, however, taken its toll. His father's earlier worries seemed well grounded when, in December 1851, William fell ill, suffering from what seems to have been nervous exhaustion. James Dawson explained to Joseph Howe that the doctors were "well aware of his [William's] ardent temperament and the dangers of indulging it at present." Although his son had visited the entire province the last summer, said James, he could not now be permitted even to think about work. James offered to substitute for his son, believing William's notes would be intelligible only to him. He also suggested that Howe appoint another man to the job in a temporary capacity.[29]

Once Dawson had recovered from his illness, he told Howe that his health, family, and private affairs would not allow him to continue as before. He suggested that, at this point, more might be accomplished by circulating information, providing proper forms, and collecting statistics, than by persevering in the imperfect system of school meetings and inspections.[30] Dawson tried to resign from the position in 1852, but changed his mind. Finally, a year later, his resignation was accepted when the plan for a public school system was defeated. Dawson did, however, note the establishment of a provincial normal school at Truro, in 1855, as the greatest accomplishment of his three-year tenure as superintendent.

Dawson's influence on the educational system of Nova Scotia was so profound that one author suggests that its history simply be divided into periods "before and after" Dawson's pioneering work.[31] After Dawson's resignation, his work as superintendent was continued by two inspectors who divided the province between them. But however arduous Dawson's work as an educational administrator in Nova Scotia may have been, it was to assume a central importance *vis-à-vis* his future career. Indeed, it led, in 1853, to his appointment by Edmund Head, then Lieutenant Governor of New Brunswick, to the Ryerson Royal Commission on Education, the principal mandate of which was to examine King's College, Fredericton. It also highlighted him as the ideal candidate to rescue the foundering McGill College, the task he was to undertake when he became the institution's principal in 1855.

THE LENGTH AND BREADTH OF NOVA SCOTIA: GEOLOGICAL ASPECTS OF AN EDUCATIONAL MISSION

As Dawson himself states in his autobiography: "A continuous thread of geological observation and discovery extended through my educational work."[32] In other words, his job as superintendent of education

– particularly his tours of the Nova Scotia countryside – permitted him to conduct scientific investigations on the side. He wrote to Margaret about how he collected seaweed, shells, and minerals, all the time observing the contours and formations of the countryside, as he travelled the length and breadth of Nova Scotia. Once he even used the tin case that had held his lunch cake and figs to carry some newly acquired specimens.[33]

Some of Dawson's most important scientific work on plants and animals of the coal measures – the discovery of a skeleton fragment from the earliest North American Carboniferous reptile or batracian (*Dendrerpeton acadianum*, the generic name "concocted" by Richard Owen to mean tree lizard)[34], of the oldest land snail (*Pupa vetusta*), and of the oldest millipede (*Xylobius sigillariae*) – occurred during his busiest years as superintendent. During that period, he still found time to scour the erect tree trunks of the Joggins with Charles Lyell upon his return trip to North America in 1852.[35] Lyell bragged that they had adapted the method of American sportsmen and "tried our game;" as a result, they had "opened a new chapter" in North American geology.[36] For his part, Lyell coveted this opportunity to view the Joggins with fresh eyes and to spend several days there alone with Dawson.[37] This second collaboration resulted in a series of papers for the Geological Society of London, on the coalfields of the South Joggins and Albion Mines. Lyell urged Dawson to continue his investigation of the coastal area, where he was sure to make more "capital" discoveries.[38]

MINING

Certainly one fruitful and ultimately lucrative realm of discovery relating to Nova Scotia geology resided in the mineralogical resources indicated by its stratigraphy and fossil record. As one mining entrepreneur said to Dawson, where else but in Nova Scotia was nature "so lavish of her Treasures as to throw together so rich and extensive a deposit of mineral wealth and render it so accessible to man's wants and enterprise."[39]

Until 1858, the General Mining Association (GMA) of London exercised a monopoly over the exploitation of the mineral resources of Nova Scotia.[40] One-time director Richard Brown contended that the monopoly brought a range of economic benefits to Nova Scotia, as well as employment to scores of men. Farmers, he said, received a market for their produce; merchants found quick sale for their goods; and commercial shippers obtained lucrative freights.[41] Others condemned the monopoly, however, for removing economic incentives and discouraging exploration on the part of local residents.⁴² Dawson, himself, criticized the British capitalists as "slow to understand, unless everything be brought within the sphere of their ordinary experiences."⁴³

Whatever the evaluation of its motivation and ultimate effect on the development of the province, the GMA decided in 1848 to hire a native Nova Scotian to conduct a geological survey of Cape Breton. They chose Dawson, who, in his view, was given "the best field of action you have to offer to a geologist."⁴⁴ Despite earlier reservations, he was also to forge longlasting friendships with GMA officials who, like Dawson, cultivated scientific pursuits jointly with their mercenary concerns. Richard Brown, for example, established a reputation for elucidating the geological structure around Sydney.⁴⁵ He sent Dawson an extensive collection of fossil plants, attended Lyell's Royal Institution lecture on erect trees of the Joggins upon his return to London, and published in the *Journal of the Geological Society* beginning in 1853. One paradoxical effect of the GMA monopoly was that "esoteric intellectual and scientific concerns" dominated early geological exploration in Nova Scotia, unlike the later obsession with utility and moneymaking.⁴⁶

Partially as a result of his connections with Lyell and Logan, Dawson garnered employment to carry out assays and evaluations of coal and iron deposits for the provincial government, as well as for small mining companies and entrepreneurs, including Charles D. Archibald, Henry S. Poole, and W.J. Ross. He undertook prospecting tours for both coal and copper during the late 1840s, including the only geological exploration sponsored by the province (of the Caribou coalfield) in 1848.⁴⁷ As well, he examined the Londonderry iron deposits, the sole natural resource excluded from the GMA monopoly.⁴⁸ He was well paid for these activities: he asked for £150 for three months of field work, as well as an additional £50 to hire casual labourers.⁴⁹ Dawson's work as a mining consultant gave him not only detailed knowledge of mineral resources, but also a host of personal contacts. It would serve him well once the monopoly of the General Mining Association expired, even though Dawson by then had moved to Montreal.

THE PUBLICATION OF *ACADIAN GEOLOGY*

Dawson's investigations into the geology and mineral resources of Nova Scotia culminated in the publication of *Acadian Geology,* widely regarded as his *magnum opus*. He dedicated the book to his mentor Lyell with deep gratitude from this "young naturalist labouring in a comparatively remote and isolated position." But the book also seemed to function as a *memento mori* to his mother, who had died in his arms just a year before.⁵⁰

Acadian Geology, or "an account of the geological structure and mineral resources of Nova Scotia, and portions of the neighbouring provinces of British America," was published simultaneously in Edinburgh, London, and Pictou in 1855, although the firm of Oliver and Boyd handled the actual printing in Edinburgh. Dawson exercised a strong hand in all publishing arrangements, undoubtedly based on the knowledge he had gleaned from his father's bookselling operations.[51] Whereas the London publishers Simpkin, Marshall, and Co. automatically distributed all of Oliver and Boyd's books in England (thereby annulling Dawson's personal preference for the Longman firm),[52] the Dawson bookselling establishment in Pictou handled the North American market. The more experienced science publisher, John Murray, refused to consider the manuscript in the first place, despite Lyell's entreaties, for he felt that no book on Nova Scotia would sell. As Lyell explained to Dawson, neither massive immigration nor the discovery of gold had made Nova Scotia known to British audiences; furthermore, publishers considered fashion first and foremost. (Murray had even dared to berate Lyell on an earlier occasion for devoting too much attention to geology and too little to "things and people in general.")[53]

Dawson persuaded Oliver and Boyd to allow him to undertake the work himself, assuming all financial risk in the matter and underwriting the range of production costs, even the advertising expenses. The publishers opted for a generous press run of 1050 copies, according to an invoice supplied by the Edinburgh printer and engraver, William Home Lizars (notable for having been the first to undertake production of Audubon's *Birds of America*).[54] Naturally, Dawson was concerned about the appearance of the volume; after long deliberations and discussion of samples from abroad, he decided that half of the books should be bound in brown cloth and the other half in green, in order "to suit different tastes." Over Oliver and Boyd's objections, he wanted a small, portable format similar to that of James Nicol's *Geology*;[55] he also came up with the idea of stamping the spine with a cut of the coal formation foliage in gold. He fussed about the quality of the paper, the strength of the sewing and binding, and the neatness of the embossed cover. Dawson shared his publishers' opinion that the sale price could be increased slightly to offset the expense of these improvements.[56]

In early July 1855, Oliver and Boyd sent word that the book had been published, and that they had already sold a dozen copies to Edinburgh booksellers. As agreed, 500 copies were dispatched to Pictou. For his part, Dawson recruited his friends to find subscribers among the "bigwigs and literati" in the Maritimes. He must have been successful, for by August another 200 copies were bound and shipped from Edinburgh to Pictou. Clearly, Oliver and Boyd should have been pleased with the publishing venture: by mid-February, their inventory consisted of only 39 copies on hand and 225 unbound (left in unfolded and uncut sheets), although they had actually sold only 50 copies in Britain.[57]

The portable octavo-sized text – "so small in its dimensions when compared with the later editions" – ran to nearly 400 pages.[58] The book began with an explanation of the derivation of the term "Acadian," a section that, with its exegesis on Micmac Indian etymology, tickled the fancy of virtually every reviewer. More importantly, Dawson then gave his reasons for devoting a treatise to the geology of this region. He argued that the geological structure of Nova Scotia "as exposed in its excellent coast sections" provided a "key" to the geology of adjoining areas. Moreover, he said, Acadian geology functioned as a discrete and unique geological district "distinguished from all the neighbouring parts of America by the enormous and remarkable development within it of rocks of the Carboniferous and New Red Sandstone systems."[59]

Surprisingly, Dawson continued, although the province's mineral resources sector had been extensively developed by private mining companies and its structure "somewhat minutely examined," it (unlike that of most of the rest of North America) had not benefited from the largesse of a government-supported geological survey. Dawson traced that absence to a complex set of political and historical reasons, particularly to Nova Scotia's low status as "one of the more obscure and insignificant dependencies of the British crown."[60]

Despite the lack of a public survey, Dawson explained, the extraordinary geological features of the province had attracted eminent geologists from abroad, while inspiring its native sons. The result was an extensive yet scattered literature on the geology of Nova Scotia. In *Acadian Geology*, Dawson provided an historical account of these endeavours, seeking to unite the salient points of these studies and supplementing them with his notes and observations collected over the last fourteen years. Particularly epoch-making, according to Dawson, were Lyell's geological explorations of 1842, which provided insightful analyses while serving to forge a link between local observers and metropolitan naturalists. Dawson concluded the introductory chapter with a bibliography of papers published by himself and others on the geology of Nova Scotia since 1842.[61]

The introduction was followed by fourteen more chapters. Three described the geological formations of Nova Scotia in general, its alluvial deposits, and the creation of its boulders and rocks. The remaining chapters treated specific portions of its stratigraphy; a discussion of "The New Red Sandstone," for example, occupied three chapters. But the heart of the work – stretching to five chapters – dealt with Nova Scotia and New Brunswick coal or the "Carboniferous System," which was seen as "the most productive field of investigation."[62] Here Dawson treated his favourite haunts: the area around Pictou, the Joggins, Horton Bluff, the cliffs of the Shubenacadie River, and Cape Breton. He discussed mineral deposits, both coal and gypsum, but also considered the area's remarkable fossils, including reptilian remains and erect trees. The Devonian and Upper Silurian strata were treated together in the fourteenth chapter, and the final chapter dealt with the metamorphic rock of the Atlantic Coast. An appendix brought together a definitive list of Acadian fossils, derived from the studies of Lyell, Richard Brown, and James Hall of New York, as well as Dawson's own collections.

Authorities at home and abroad praised the book. In its quarterly installment of reviews of scientific books, the *Westminster Review* introduced *Acadian Geology* immediately after a review of a book by the Oxford professor of geology, John Phillips, pronouncing it "not unworthy to rank near it." The reviewer placed *Acadian Geology* in the class of books that dealt with "important yet distant regions of the globe." He applauded the book's capacity to deepen knowledge in the area, its treatment of practical issues (namely, mining), and its provision of useful evidence "in regard to problems of the highest interest to the geological speculator." Besides dealing with the derivation of the word "Acadian," much of the review treated the fieldwork of British geologists William Logan and Lyell in Nova Scotia. All in all, the reviewer felt that Dawson had succeeded well in his two aims: the writing of an elementary and popular account for lay colonial readers; and the provision of accurate, original scientific information for geologists everywhere. In short, the author praised *Acadian Geology* "both for what it actually contains, and for the rich promise of future discovery which it holds out."[63]

A shorter review in the *Athenaeum* two months later spoke mainly of the practical implications of the book. The potential promise of Nova Scotia coal made Dawson's work of immense interest "throughout the overpopulated countries of Europe, and especially of Great Britain."[64] These accolades from overseas, which were echoed by his mentor Lyell, assumed special significance for Dawson. Lyell wrote that he knew "of no modern work on the geology of any country in which the author is entitled to speak with so much authority as an observer, naturalist, chemist and mineralogist."[65] Even the paleontologist Sir Philip Egerton sent his thanks from Glasgow for the "elegant treatise."[66]

In Canada, *Acadian Geology* appeared as the lead review in the January 1856 issue of the *Canadian Journal of Industry, Science, and Art.* The reviewer, E.J. Chapman, pronounced the volume "well-timed," and particularly praised Dawson's chemical examinations of various samples of coal. Sales in Montreal became brisk when the news of Dawson's appointment as principal of McGill broke; one bookseller immediately ordered three dozen copies.[67]

The publication of *Acadian Geology* offered a loving tribute not only to the geological formations of Dawson's native province, but also to his family and friends there. A decade earlier, Dawson had described to Margaret his gratitude to, and affection for, his parents, as well as his determination "never to desert them and their service as long as they require my assistance."[68] The closeness of the family is summed up by the elder Dawson's statement that if his wife had any fault, it was in "indulging in too much affection for Near-Relations."[69] Undoubtedly, her death helped to release William from the bonds that held him so firmly to the family homestead.

William, for his part, might well have echoed his former schoolmaster McCulloch's sentiment: "It would be like tearing the flesh from my bones to leave Pictou."[70] But Dawson, like McCulloch before him, suffered the separation. Pictou had given him such a special preparation – both in terms of schooling and the remarkable world of nature that lay at his doorstep – that it became his mission in life to use these unique advantages to good purpose elsewhere.

4 A Real Horse Race[1]

Rambles through the Nova Scotia countryside and his special friendship with Charles Lyell developed Dawson's geological sensitivity, whetting his appetite for knowledge. The Pictou Academy had given him an outstanding secondary education grounded in works of the Scottish enlightenment. Still, he sought to advance his scientific understanding in a more formal way. On two occasions during the 1840s, Dawson travelled to the University of Edinburgh, an institution well known in nineteenth-century Nova Scotia, both for its distinguished alumni and its faculty members' widespread publications. Edinburgh had emerged by this time as one of the few places in Great Britain where natural history sciences were taught as part of the university curriculum, thanks largely to the labours of Robert Jameson, the Regius professor of natural history from 1804 until his death in 1854. These experiences helped groom Dawson for the race of his life.

THE STAKES[2]

Jameson had attracted both admirers and detractors over the years. Like Dawson, other aspiring naturalists (by choice) as well as future physicians (by statute) attended his classes for systematic instruction in the basics of the field. The young Charles Darwin, a critic, found Jameson's lectures so excruciatingly dull that he vowed "never ... to read a book on geology, or in any way to study the science."[3] Furthermore, the fees that Jameson marshalled from class attendance (three to four guineas per student) and the admission fee charged for the associated museum of natural history (as much as a guinea per year) were considerable, since his course was compulsory for medical students.[4]

Dawson and scores of other young men might easily have been dashed by Jameson's taciturn and reserved disposition, what one contemporary described as an "ungeniality of nature." But Jameson, unlike other naturalists of the day, had managed to establish a centre or school for the study of natural history. This came about because of his enthusiastic support for all aspects of the science, although he was strongly committed to the geological doctrines of Abraham Gottlob Werner and personally favoured the pursuit of mineralogy.[5] The creation of the Wernerian Society (in 1808) and the *Edinburgh Philosophical Journal* (in 1819, which he edited by himself from 1824 to 1854) numbered among Jameson's most important achievements in Edinburgh, although it was his professorship and museum keepership that stood at the core of his empire. Indeed, his vigorous promotion of the natural history sciences helped to secure the university's enduring auspicious position, once the landscape of Scottish enlightenment began to fade. And he himself attained an international reputation.[6]

During the course of Jameson's fifty-year tenure at Edinburgh, the natural science curriculum and its role in the university were subjected to repeated scrutiny. Long and protracted debates centred on the cultural and social functions of the Scottish universities as well. Much of this discussion and dissension resulted from the great university's apparent mid-nineteenth-century decline. That is, once the University of London and Queen's College, Belfast, were able to answer the needs of medical students, the number of students enrolling at Edinburgh fell dramatically. Accordingly, few new professorships were created in the arts faculty until the second half of the nineteenth century.[7]

The Scottish educational ideal nevertheless demanded that the university remain an integral part of the urban, and even the national, environment. What made Edinburgh so appealing to a young colonist like Dawson was its emphasis on the democratic and the practical, in contrast to the elitist, almost other-worldly mores of the Oxbridge residential colleges.[8] Mind you, critics accustomed to the English model felt that the University of Edinburgh was deteriorating as a consequence of the internal inconsistencies brought into play by these values. In their view, the enormously powerful professoriate was unbalanced by the inadequately prepared group of students, dearth of functional textbooks, and absence of a proper system of examinations. As a result, the intellectual energies of the professors seemed dissipated.[9]

Divergent philosophies of education clashed dramatically when university chairs fell vacant, usually due to the death of the incumbent, and successors had to be appointed. These contests became almost "a feature" of Scottish intellectual life during the nineteenth century.[10] Particularly acrimonious battles were waged over the chairs of moral philosophy (1820), natural philosophy (1832), and mathematics (1838). Nonetheless, historians identify a new era beginning with the 1850s, when the succession to chairs "gave rise to gestures of national cultural disunity and to remarkable outbursts of sectarian and theological animosities."[11] With fewer students and staff, competition intensified over the limited remaining resources.

A dismal record was particularly evident with regard to science chairs in the Faculty of Arts. A professorship in astronomy, created in 1785, often fell vacant, since so few students were attracted and the government failed to supply adequate instruments. When the Regius professor of technology, George Wilson, died in 1858 (three years after his position had been established), the chair was simply abolished.[12] The natural history professorship, housed in the Faculty of Medicine, however, enjoyed a more secure footing within the university, for despite declining enrolments, the medical school was to remain the most vibrant sector of the university throughout the century. Medical chairs – along with classical languages – could be counted among the most lucrative, since the curriculum in that faculty featured a succession of largely compulsory courses for its students (who paid the professor directly for lectures attended).[13] Undoubtedly, too, Jameson's fifty-year reign in the Regius chair heightened the clamour and expectations among the ever-growing group of naturalists; here must be a position worth the fight, they thought, a prize that merited the race.[14]

PRELIMINARY CANTER

Almost eerily (given the context of competition and strife), when Jameson died in 1854, the chair passed – with relatively little discussion or contest – to the Manx marine biologist, Edward Forbes, Jr. Forbes was as affable as Jameson was reserved. For more than a decade leading up to Jameson's death, throughout the long years during which he was too ill to lecture but too stubborn to retire (so "tenacious of life," as one colleague put it), Forbes had focused his energies on obtaining the chair.[15] Although he anticipated a brisk contest, few, in fact, were willing to threaten his dream.[16] Universally acclaimed for his brilliance and wit, Forbes was thought to be a fine choice by influential politicians, leading scientists, the other Edinburgh professors, and town councillors alike.[17] As one biographer states, "The universally admitted genius and range of acquirements of the new Professor disarmed all petty cavilling, and there arose one unbroken note of joy and welcome."[18] Alas, he succumbed within six months to a kidney ailment. On his deathbed, a critically ill Forbes blamed the Edinburgh town councillors for having forced him to leave London precipitously, in order to assume the chair he had sought so long. (He also had been asked to catalogue the natural history museum in the space of three months.[19]) He bitterly remarked that now "the bailies have killed the goose that laid the golden eggs."[20]

A GLIMPSE INSIDE THE PADDOCK[21]

In stark opposition to the earlier succession, the ensuing contest for Jameson's chair could not have been more intense or prolonged. Lyell opined to his protégé Dawson that the number of candidates who had been proposed by their friends or who had entered the competition voluntarily was "unprecedented in the history of this or any other chair vacant in Great Britain." He attributed the great number of candidates to the *éclat* given the position by Forbes, who quite possibly might have earned £1600 a year from the associated salary and fees. (Forbes had counted more than 130 "paying" students in his 200-seat, filled-to-capacity lecture hall.[22]) As a result of this publicity, the university and townspeople of Edinburgh had become overly ambitious and, according to Lyell, were persuaded erroneously that the chair was one of the most profitable, if not indeed the most valuable, in Europe.[23] A more skeptical professor, John Fleming, maintained, however, that the influence of the chair was declining in relative terms, and that it was unlikely to net an annual salary of £900.[24]

The candidates vying for the "El Dorado" (as Lyell described it) included the most distinguished naturalists already holding the most important positions elsewhere in Britain and abroad. One journalist quipped about the "eminence on crutches here, and of Mr So-and-so, 'who has been too long kept back,' there."[25] The geologist John Phillips, who had just assumed a chair at Oxford, declined the enthusiastic support of the Edinburgh professor of natural philosophy, James Forbes, who wrote to him just two days after Edward Forbes died.[26] The eminent London comparative anatomist, Richard Owen, complained about the journals that had declared him in the running. Lyell felt that the Swiss naturalist Louis Agassiz, by then well-established in the United States, would not consider standing for the job, despite the urging of his promoters. (Indeed, with the prospect of creating his own museum at Harvard, Agassiz had declined offers for professorships at the Paris Natural History Museum and the Zurich Polytechnical Institute around this time.)[27]

One may well wonder what might have induced a figure of international reputation to relocate to Edinburgh. Edward Forbes claimed to have found much freedom in Edinburgh, compared to the many demands on his time in London. He had assured Edinburgh's detractors that "those who talk of banishment forget how many clever men are here still."[28] James Forbes contended (to Phillips) that Edinburgh probably outshone Oxford in terms of emolument associated with the chair. Moreover, he argued, an important museum, about to be further enriched by the spoils of the geological survey, was attached. Edinburgh housed a distinguished Royal Society, and professors enjoyed a "generally agreeable position" about town.[29] Thomas Henry Huxley – who seemed "marked out" as Forbes' successor at Edinburgh, as had also occurred at the School of Mines in London – thought the enticement of a salary in the range of £1000 was "not to be pooh-poohed," although he felt that London "is *the* place, the centre of the world," preferable "to half a dozen Edinburgh chairs."[30]

THE STARTERS

If the first-rank naturalists did not genuinely consider taking the job, a whole group of qualified, but lesser known, amateur naturalists were in hot pursuit. Both Thomas Stewart Traill (who already occupied the chair of medical jurisprudence at Edinburgh, and had often lectured on natural history for Jameson) and John Fleming could be counted among them, although as septuagenarians their claims were weakened (Forbes had been thirty-nine.) Another Scottish landed proprietor, Sir William Jardine (who had hosted Forbes on several occasions at his country estate in Dumfries county), had entered the contest but, it was felt, with little chance of success.[31] Gossip about the succession began to reach a feverish pitch: the *Scottish Press* compared the process to the "mysterious and suggestive winks and shrugs and pokes of the elbow wherewith the knowing ones hint the favourites and the state of the odds at the betting houses the night before the Derby."[32]

If a suitable candidate to cover the entire range of subjects could not be found, one solution would be to divide the chair into zoological and geological components. (Earlier Jameson had proposed that the museum be split off from the chair, and that the collections obtain a special superintendent; namely, his nephew, Laurence Jameson.[33]) This eventuality was unlikely, however, for no special endowment for the geological, mineralogical, and paleontological portion could be supplied. Medical students, who already faced enough compulsory lectures, might then have been required to attend only the zoological part, leaving geology virtually deserted. In the remote case of a divided chair, Hugh Miller, an enormously popular local figure who had occasionally audited Forbes's lectures, could lay claim to the geological portion; Huxley, the zoological. But, as Lyell commented, this resolution of the problem – reflected in a proposal by the Town Council to divide the chair – made everyone "feel at sea."[34]

Candidates were forced to lobby on a variety of political levels. Officially, the Regius professorship was a Crown appointment at the pleasure of the monarch; in fact, the name of the appropriate individual was passed to the queen from the secretary of state for the Home Department. For matters concerning Scotland, however, the secretary deferred to the opinion of the lord advocate of Scotland, dubbed by one historian "a patronage broker for the government north of the border."[35] Accordingly, lobbyists seem to have focused their energies on influencing the opinion of the lord advocate, Sir James Wellwood Moncrieff.

In this particular case, the lord advocate acted in close collusion with the Town Council. Historically, the Town Council (as the patrons of the university) exerted considerable power over university appointments. The councillors, alas, were seen even by the Scottish-born Lyell as a parochial clique, responsive only to social pressures of the day as defined by them. For their part, the councillors had been embroiled in what has been termed a "Thirty Years' War" with the Academic Senate of the University. This long-standing dispute would not be resolved for another four years, at which time regular university appointments were placed under the wing of a Board of Curators. Although this development would effectively deprive the Town Council of their strong hand in university affairs, their power in 1855 had not yet been checked.[36]

Political struggles were further complicated by religious considerations. The former test act, obliging lay professors to accept the *Confession of Faith* and pledge allegiance to the Church of Scotland, had just been abolished, in 1853.[37] But this did not mean that candidates of all faiths were treated equally, although now Free Churchmen, along with Episcopalians, could gain access to university chairs. (Indeed, one historian argues that the act of 1853 only increased a dangerous tendency, whereby electors to chairs preferred a less qualified candidate of their own denomination to a better qualified one of another.)[38] Lyell protested, for example, that William B. Carpenter – a man of "irreproachable character and high moral worth" – had had his Unitarianism weigh against him. Similarly, Hugh Miller's "free-Kirkism" would prove an obstacle, as did Agassiz's "want of orthodoxy."[39] Lyell concluded that "in an age of theological bigotry a man had better declare nothing than belong to an unpopular sect if he wants to turn his talents to account in the market," especially among "the narrowminded in the north."⁴⁰ Clearly, Lyell had little regard for the cultural values of the inhabitants of his ancestral home, and his suspicions about their petty motivations would prove to be well founded.

ENTER DAWSON AS A DARK HORSE

Just four days after Forbes's funeral on 23 November 1854, Lyell approached Dawson about standing for the chair. It was still a "great lottery," said Lyell; no one was assured of walking away with the honour. The most highly acclaimed for the position appeared disinterested; the second-ranking candidates all had obvious shortcomings. As Lyell summed up the field: "One is too old, another not a good lecturer, another a good zoologist but quite ignorant ... of mineralogy and geology etc."⁴¹

Lyell, who had earlier been a significant force behind Forbes's quest for the Edinburgh chair,⁴² saw Dawson's great advantage as "his youth taken with reference to his attainments." He intended to argue, he explained, that Dawson's love of science and present qualifications would enable him to surpass those older, middle-aged scientists who, at present, could claim more publications.⁴³ For Dawson – who so quickly accepted Lyell's offer of support that he sent a telegram via New York – this was a position about which he had only dreamed when it was last vacant.⁴⁴ He worried about his weakness in British zoology, where Forbes had been especially strong. Generally, however, he believed that he had studied a broader range of subjects than most naturalists, including mineralogy and geology, as well as general and North American zoology.⁴⁵

Over the next two weeks, Dawson had the opportunity to reflect more soberly on the Edinburgh position. He wondered whether the "scramble" for the chair was not based on an unrealistic expectation of its value; surely, without a large influx of students into the university, its annual remuneration could not exceed £1000. He observed that the proposal to split the chair into zoology and geology was ill conceived, believing that such "narrow specialization is harmful to good science." By habitually leaving his course unfinished, said Dawson, Jameson had (perhaps unwittingly) given ammunition to the opponents of an undivided chair; yet this was not due to the field's intrinsic extensiveness, but rather to Jameson's giving "undue time to details and favourite hobbies." Dawson outlined how he would structure the course to cover the subject matter, although he felt his chances were hurt by his "present obscure position." In any event, Lyell was to act as the clearing-house for Dawson's application and testimonials; afterwards, he intended to deposit the bundle with the home secretary, Lord Palmerston.[46]

Lyell warned Dawson to emphasize his own achievements and not to mention his name. He suggested that Dawson's father-in-law, Leonard Horner, could wield influence in Edinburgh circles on Dawson's behalf, but that a recommendation from their mutual acquaintance, Edmund Head, then governor-general of Canada, was essential. Lyell also advised that testimonials from the Duke of Argyll, the botanist William J. Hooker, the Nova Scotia government, and the mining companies that had employed Dawson would be effective. Certainly, he added, Dawson should mention matriculation at the University of Edinburgh and, in particular, attendance at Jameson's classes for several semesters. The support of American scientists, on the other hand, might prove to be a liability, as it would tend to make him appear American rather than British.[47] Even if Dawson did not get the job, insisted Lyell, it would do him no harm for future contests to collect testimonials and make his name better known in Britain.[48] Another supporter concurred with this assessment, and added that "in great attempts it's glorious even to fail."[49]

Dawson followed Lyell's directives to the letter. At home, he approached several clergymen, including Alexander Forrester (an alumnus of the University of Edinburgh and Dawson's successor as superintendent of education for Nova Scotia); his own minister, Andrew W. Herdman of the Church of Scotland; and James Bayne, another Presbyterian clergyman. William Logan, now writing with authority as director of the Geological Survey of Canada, said that Dawson's special aptitude for the job would be to bring an accurate appreciation of North American geology to Europe. The secretary of the Pictou Academy, William James Anderson, attested that Dawson had delivered thirty-two lectures on natural history there during the academic year 1848-49.

Several Nova Scotia politicians provided testimonials, including William Young (attorney general), Sir Gaspard Le Marchant (lieutenant-governor), and Joseph Howe (recently named provincial secretary), who referred to Dawson's "gentle manners" and "sterling integrity," as well as his "remarkably attractive style as a lecturer." He found support from mining entrepreneurs, including Henry Poole, former manager of the Albion Coal Mines, and Richard Brown of the General Mining Association of London. Abroad, Dawson felt he was not without influence, for his wife was "an Edinburgh lady," with "a respectable circle of relatives and friends there."[50] He also wrote to the Edinburgh publisher William Chambers, as well as the Edinburgh professors William Gregory (with whom he had studied chemical analysis in 1846-47) and James Forbes.[51]

Dawson's official letter of application for the chair shows how closely he followed Lyell's advice, yet it also provides a revealing glimpse of how the thirty-five-year-old Canadian scientist saw himself. He chose to present his credentials in light of the disadvantages of his personal circumstances, and even claimed that the attainment of the post would enable him to encourage others "in isolated positions and remote parts of the world" to overcome similar obstacles. Nevertheless, he recounted that when he first came to Edinburgh during the winter of 1840-41 to attend Jameson's courses, he found himself better prepared than most of his peers. Not only had he already studied natural philosophy and chemistry at Pictou Academy, but he enjoyed considerable experience as a draftsman, taxidermist, and geological collector. He explained that he had subsequently returned to Edinburgh for the 1846-47 academic term, in order to study chemical analysis in connection with his work as an assayer for several mining companies in Nova Scotia.

Since that time, continued Dawson, he had been employed as an educational superintendent and consultant for Nova Scotia and New Brunswick, although family responsibilities had forced him to resign from these posts. His papers published during these years, said Dawson, had been written "in circumstances which debarred me from access to libraries of reference and public collections, and therefore under greater disadvantages as compared with naturalists resident in Great Britain."[52] Clearly, Dawson wished to present himself as a colonial who was handicapped by his situation but not unqualified. Given resources comparable to those at the disposal of his competitors, he would – in his own and Lyell's view – leave those competitors in the dust.

DELAY AT THE POST

James Forbes's early prediction – that the vacant chair would have to be filled soon after Edward Forbes's death – was clearly off the mark. By the end of January, Lyell was no longer so sanguine about Dawson's chances of success; every day brought news of another formidable rival. He had never seen such a competition, he wrote; some half-dozen eminent men "almost of European reputation," along with dozens of respectable candidates (such as Sir William Jardine's close friend, George Johnston of Berwick, a physician who worked as easily in botany as in zoology) were doomed to disappointment.[53] Scores of other candidates jockeyed for position. Lyell discouraged Dawson from making a personal appearance in Edinburgh, once he learned that men such as Owen, Phillips, and Agassiz had been invited.[54] By early February, he reported that Agassiz had apparently declined by letter, but that his supporters had not given him up entirely. But whether this was due to their tenacity or to "some reserve" in Agassiz's letter, he could not say. Huxley – who seemed to have the best chance of all the candidates – had, said Lyell, been "bribed by the Government here not to go to Edinburgh" but to remain in London. Now, added Lyell, he often heard mention (especially by the medical people) of the name George Allman, a Dublin professor.[55]

Dawson, however, was not about to leave a stone unturned, as one supporter had so exhorted him.[56] If he did decide to visit Edinburgh, Lyell advised him to deliver a lecture, as "candidates for a kirk give sermons to show how they can preach."[57] (In the absence of any conventions for displaying the qualifications for a professorship, the closest analogy that sprang to Lyell's mind came from the ecclesiastical realm.)[58] Although hindsight might argue differently, Dawson's promoters maintained that he should not undertake the expensive trip; if he went and did not conquer, it would be forever "a source of mortification" to him.[59] Lyell believed, moreover, that no one was seen in his best light when "on probation."[60]

As events unfolded, Lyell's view – that the government was resolved not to hurry the appointment – proved to be an understatement.[61] Rumour had it that the home secretary, Lord Palmerston, was so overwhelmed by his correspondence, that he had twenty large boxes constructed to accommodate unanswered and unopened letters. (He was reported once to have found a two-month old letter that begged him to order a clergyman to bury an unbaptized corpse!) Lyell maintained that Palmerston would not, in the end, read a single certificate, but would refer the whole matter to his subordinate, the lord advocate of Scotland.[62]

Just when Lyell was satisfied that he could prevail upon all interested parties, whether by dint of personal friendship or by vigorous lobbying, a ministerial reshuffling occurred, changing the portfolios of the principal actors. Palmerston left the Home Office to become premier; another acquaintance of Lyell's, Sir George Grey, became secretary. Subsequently, however, yet another government crisis intervened, and Lyell worried that even Grey might be out of office.

Political and personal accident can doom even the best-laid strategy. As Lyell reflected several months later, "In the chapter of accidents there is no saying what may turn up."[63] Lyell announced that he would be absent on the continent for the next six weeks (during March and April), and would place all of Dawson's documents in the capable hands of Geological Survey of Great Britain director Henry De La Beche. To the latter, Lyell explained that Dawson's attainments were so remarkable – especially in the domain of fossil botany – that even Hooker had been impressed. No individual, added Lyell, could be expected to shine in all the branches represented by the Edinburgh chair, especially someone like Dawson, only in his mid-thirties.[64] Nevertheless, De La Beche could hardly have been expected to press the candidacy of a stranger with the same enthusiasm as that exhibited by Lyell for his protégé.[65]

De La Beche countered that Dawson might be interested in working for him. He called the Edinburgh chair a "false position" and tried to entice Dawson to work on the geological survey in Scotland. He complained that his best men were constantly being lured away to work in India and Australia; for example, he had just lost one geologist to South Africa, where he expected to earn £1000. Although he could offer Dawson only between £200 and £250, fringe benefits included working with the survey corps in the field, using their museum, and being able to publish one's own original work (including illustrations) at national expense. In support of De La Beche's efforts, Lyell observed that the position might serve Dawson as a stepping stone to other posts in Britain.[66]

From Dawson's perspective, this was the worst possible time for Lyell to be absent from Britain, for it was precisely the moment when George Allman chose to launch an offensive. In early March, Allman sent an enlarged version of his printed testimonials to the lord advocate. It included attestations from the Paris zoologists Henri Milne-Edwards and Armand Quatrefages, the Oxford professor John Phillips, Charles Darwin, and several distinguished Irish luminaries. In the words of another candidate, "The Edinburgh Baillies have great faith in such things."[67] But perhaps his ace was an earlier letter of recommendation from Edward Forbes.[68]

TIPS FROM AN INSIDER

At this point, Peter Bell, Dawson's brother-in-law, decided to try, himself, to advance Dawson's claims in Edinburgh. Having learned that an offer to Agassiz was imminent, he persuaded a friend, the editor of the *Scottish Press*, to run an article comparing Dawson and Agassiz, to the detriment of the latter. This, he crowed, "fairly launched" Dawson before the public. He sent copies to Lyell, geologist Sir Roderick Murchison, and each Edinburgh professor.[69]

Lyell, however, decried this "illiberal attack on Agassiz," who had no intention of taking the chair. It was all "too well suited to the prejudices" of several Edinburgh town fathers, said Lyell.[70] He also felt that it would "cool the courage" of Dawson's supporters to portray him as an "out and out Presbyterian" who would "introduce references to the Bible when fitting occasions occurred in his natural history lectures."[71] After Dawson disassociated himself from the campaign against Agassiz, Lyell conceded that he was not answerable for the "injudiciousness" of his friends, whose zeal and activity had been otherwise great.[72] As well, Dawson assured Lyell that the article had been written with the best intentions, although he wished that his name had been put forward without disparaging Agassiz's. Dawson felt it was inevitable, however, that the Swiss naturalist would be attacked for his views on man's creation, which (like his glacier theory) did not take sufficient account of particular evidence.[73]

Bell also arranged to have over a hundred copies of Dawson's testimonials printed, and dispatched them to everyone of influence: professors, town councillors, politicians, aristocrats, writers, and publishers alike. At least now, he congratulated himself, Dawson and his qualifications would be known far and wide, whereas many in Edinburgh had never heard of him until three months ago. A group of Edinburgh professors assured Bell that if Allman won the chair, Dawson would "very likely" get Allman's position in Dublin. In the meantime, Bell urged Dawson to mobilize any support he had in London "to bore the Government, and not leave a stone unturned." Bell was certain that "this stir is sure to produce something."[74]

A fortnight later, Bell reported that the chair still stood vacant, and that the Town Council had prevailed upon Traill to deliver the summer course in natural history.[75] He also relayed the news that De La Beche – who had been in the midst of assembling Dawson's portfolio of certificates and attestations – had suddenly died. Bell used even this as an opportunity, asking his editor friend to "very neatly dovetail" Dawson's and Lyell's names into De La Beche's obituary in the *Scottish Press*. Bell believed that the most immediate consequence of this event would be Murchison's (De La Beche's successor as survey director) throwing of his support behind the Regius professor of geology at Aberdeen, James Nicol. If Nicol went to Edinburgh, reasoned Bell, Dawson might then be named to the Aberdeen chair. Bell concluded that if either Nicol or Allman won the Edinburgh professorship, Dawson should be prepared for "an energetic attack on the government" in order to obtain the vacated chair, whether at Dublin or Aberdeen.[76]

FIGHTING TO THE FINISH

By the end of May, the chair was still vacant, although Allman appeared to have moved out in front of the competition. Not only had he favourably impressed the Edinburgh professors during a recent visit, he had also won the critical support of Richard Owen. Nevertheless the government was still "dilly-dallying," and talking of splitting the chair between two native sons: Hugh Miller and John Fleming.[77] For his part, Allman was just as anxious as Dawson; he asked Huxley if he could "penetrate the mystery" surrounding the Edinburgh chair, as he found the delay "harassing."[78]

By early July, Lyell reported that Dawson and Allman were the two remaining candidates under consideration. Time was running out (which Lyell felt might promote Allman's case, given that he resided in Britain) and summary arguments were being made about the candidates' respective strengths. Irish geologists, especially Joseph Beete Jukes, exaggerated Allman's paleontological accomplishments, while his defenders minimized Dawson's ability to teach zoology, especially comparative anatomy.[79] Finally, after more than half a year of promoting Dawson's interests, Lyell decided to rest his case. He explained to Bell, that with the whole matter before the government, he could no longer "with propriety interfere."[80]

What Lyell did not reveal to Bell (or to Dawson, for that matter) was that he had just come from speaking to the home secretary and the lord advocate about the chair. Apparently both he and Murchison had been told that Allman was to be appointed soon. In the view of the two senior geologists, the botanist Allman's success meant that "those branches of science in which we are more specially concerned and on which Edinburgh has hitherto lent us such a helping hand should, in future, be less favoured."[81] Lyell saw the work of the Duke of Argyll in the decision for Allman, and mentioned Murchison's contention that "had Dawson been personally known in this country – had you been able *to trot him out* – he would have stood a very different chance." Indeed, Lyell admitted that perhaps he had erred in not encouraging Dawson, an even better speaker than a writer, to make the trip.[82]

Lyell also had his father-in-law, Leonard Horner, interrogate one of the distinguished Edinburgh medical professors, William P. Alison, about the position. Although this was perhaps a tactic to win Alison's support, it was more likely an attempt by Lyell to learn whether a geology chair would be created in the near future. The correspondence suggests that the indecision over the division of the natural history chair (into zoology and geology) had largely contributed to the delay in making an appointment.[83]

Horner's questions to Alison were designed to cast Dawson in a stronger light than Allman, while never leaving the realm of abstraction. He asked whether zoology (including comparative anatomy) was especially emphasized in student examinations in natural history, and whether the course was not compulsory for other than "direct professional objects." That is, queried Horner, was it not true that natural history was included in the curriculum "for the sake of liberalizing the mind of the student," thereby introducing him to a wider range of subjects than those required in other medical courses.[84]

Alison responded that, indeed, natural history was obligatory for its liberalizing tendencies. Nevertheless, he contended that the chair should be left intact because, given the present state of enrolments in the medical faculty at Edinburgh, two professors could expect no more than £350 each in fees and salary should the division occur. From a scientific perspective, however, Alison maintained that a division into "inorganic nature" and "animal life" would be desirable, as both subjects had been "extended, magnified and subdivided" during the last fifty years. He argued that to attract "a man of such eminence as we ought to have" would require a separate endowment for geology and allied sciences, which "could do more for the character and usefulness of our University, than many thousands spent on stone and lime, or even on books."[85]

Lyell felt that there was no reason to assume that such a division would occur in the near future. Bell, for his part, disputed Lyell's assessment, not knowing of his work behind the scenes. He believed that Allman's support was confined to a few professors and that, with Agassiz out of the picture, Dawson should consider making a personal appearance in Edinburgh. According to Bell's reconnoitering, Dawson's biggest liability was that he was labelled a "Lyellite" – someone who endorsed all of Lyell's opinions. In Bell's opinion, the best tack for Dawson at this point would be to meet his opponents; that would surely win them over.[86]

BEATEN BY A HEAD

A personal appearance by the apparently charismatic young Dawson would not, however, come to pass; two weeks later Bell sent him word that Allman had won the chair.[87] The university was undergoing curricular changes that were to transform the chair into a zoology position with a very minimal role for geology. According to the terms of Allman's appointment, the university reserved the possibility of removing both geology and mineralogy from Allman's purview at some future date, whenever a decision was made to reduce the chair's range of responsibilities. At least, Lyell comforted himself, Allman would "inocculate" his students with a taste for geology.[88]

An article in the *Scottish Press* analysed the appointment, after having run a piece promoting the interests of the "distant colonist" unfairly "pitted against candidates who are on the spot and who had each his circle of supporters around him." It stated that despite the ways in which Dawson's detractors had tried to minimize his suitability for the position, he had been a strong contender, perhaps even the leading candidate for the job. Nevertheless, once Murchison began to press for the appointment of James Nicol, Home Secretary Sir George Grey decided that he could not afford to offend either him or Lyell. Accordingly, he settled on the third, if less well-qualified candidate, Allman.[89] In a less analytical vein, the *Scotsman* simply quipped that at least Allman's knowledge of geology would be improved.[90]

The contest was not without benefit to Dawson; not only had he achieved a high ranking on the list of candidates, he had also made his name better known in Britain.[91] In August, Lyell proudly explained to Huxley that his "friend Dawson" had been offered the direction of McGill University, "the principal medical college in Canada."[92] In September, Dawson decided to attend the British Association meeting in Glasgow, where Lyell sat as vice-president. He read a paper on paleontology to a large audience, including "all the magnates": the Duke of Argyll, Murchison, Adam Sedgwick, Hugh Miller, and Lyell. There, Lyell told him that if it had been read the year before, he might well have gotten the Edinburgh chair.[93] George Wilson, the first incumbent of a recently created professorship in technology at Edinburgh, agreed that a personal appearance might have won Dawson Allman's post, which he characterized as having gone "a-begging."[94]

Indeed, a series of "near-misses" – the kind of accidents that befall even the best racehorses – would seem to have attended Dawson's unsuccessful attempt to obtain the chair. Besides the already described circumstances beyond his control – such as the untimely death of De La Beche – other unforeseen turns of fate seemed to conspire to keep Dawson in Canada. The *Acadian Geology*, for one, appeared too late to advance his interests, and even the Edinburgh publishers Oliver & Boyd declined to intervene.[95] Around that same period, Dawson learned that he had been named a Fellow of the Geological Society; one might well contrast this achievement with that of Allman, eight years Dawson's senior, who had just been elected a Fellow of the Royal Society.

As the race for the Edinburgh chair drew to a close, it became apparent that Dawson's luck was about to change. The seeds of subsequent victories were sown in this particular defeat; already it seemed more instructive than crushing. Years later, Dawson, looking back on these events, reflected that he was still astonished at his temerity but that "youth and hope" had counted for much.[96] Indeed, his hopefulness was shortly to be rewarded. Fate would soon smile on young Dawson and on all his undertakings.

5 "Good results in store"[1]

News of his defeat in Edinburgh reached Dawson in the autumn of 1855, just as he was about to embark for the British Association for the Advancement of Science meeting in Glasgow. In Halifax, awaiting the departure of the steamer for Liverpool, Dawson received the telegram that "dashed all my hopes."[2] But at virtually the same moment, word came that he had been offered the principalship of the University of McGill College in Montreal, upon the recommendation of Sir Edmund Head.[3] As Dawson relates in his autobiography, he appeared in Scotland, much to his own surprise, not as "a candidate for the Edinburgh chair," but as "the principal-elect of McGill," seeking there a university degree at the request of the Board of Governors.

On his return to Nova Scotia soil, Dawson joined his wife and three young children, George, Anna, and William Bell, and, together, the family set off to begin a new life in Montreal.[4] Perhaps the words of Joseph Howe were echoing in Dawson's head as he departed for parts unknown: "This country wants laborers who are neither parsons nor lawyers. You are one of them."[5]

How remarkable that the thirty-five-year-old Dawson was willing to leave Nova Scotia for a part of Canada that he had heard little about and never seen. Just a few years before, he had refused to move from Pictou to Halifax, despite his appointment as education superintendent for the province. Surely, Montreal – a cultural, religious, and linguistic Babylon of around 100,000 souls – would have appeared considerably more foreign, if not forbidding, to someone accustomed to the relatively homogeneous population of Pictou. As he relates in his autobiography, he had fewer friends in Montreal than in Edinburgh, and the move meant, or so he believed at the time, that he would have to abandon his research into Carboniferous and Devonian paleobotany.[6]

Montreal nonetheless presented a number of advantages and challenges to Dawson. By the 1850s, it had begun to emerge from the ranks of provincial towns to become the tenth largest city in North America.[7] Certainly, it was "the chief city of British North America," as Dawson himself stated in his inaugural address.[8] The offer to direct a college there, however modest, must have helped to heal his wounds after the Edinburgh affair. As Dawson admits in his autobiography, under other circumstances he might not have accepted McGill's offer, but the Edinburgh experience had familiarized him with the idea of change and he already felt "partially uprooted."[9] As well, his mother's death just a year before had relieved him of the heaviest part of his filial burdens; only his father James was left behind in Pictou. How keenly Dawson felt even this separation is revealed by the affectionate letters regularly exchanged with his father during his early years in Montreal.

Cold reality abruptly shattered Dawson's sanguine expectations. Accustomed by now to the veneer of civilization associated with institutions of learning such as the University of Edinburgh, Dalhousie College, and even Pictou Academy, Dawson was shocked by the sight that greeted him in Montreal. There, McGill's campus consisted of "two blocks of unfinished and partly ruinous buildings, standing amid a wilderness of excavators' and masons' rubbish, overgrown with weeds and bushes. The grounds were unfenced, and pastured at will, by herds of cattle ... The only access from the town was by a circuitous and ungraded cart track, almost impassable at night."[10] The university possessed no library, no museum, and only a few pieces of old-fashioned scientific apparatus.[11]

Perhaps to compensate for its physical shortcomings, McGill gave Dawson an impressive salary of $2000 (£500) (increased to $2500 in 1861) and a residence.[12] The Dawson family was lodged in the front part of the east wing of the Arts Building; the back was occupied by McGill's registrar, bursar, and secretary, William Baynes and his family. (The Dawson and Baynes families would become lifelong friends.) Margaret was cheered to see the "large square solid-looking stone house," but inside a terrible state of dilapidation and confusion confronted her: "torn wall papers, cracked ceilings, squeaky floors and rat holes in a number of places." Only temporarily overwhelmed by the sight (and by the fact that the previous principal had not yet vacated the premises), Margaret set about applying her customary touch that, in the words of her granddaughter, always seemed "to turn the dull and grey into bright and burnished gold."[13]

Even amid mid-November's chill, Dawson described the family's domestic arrangements as "comfortable." The university had paid for much of the cost of the succession of "masons, plasterers, carpenters, glaziers, painters, paperers, and woodcutters" who had helped to improve the residence. Included was a woodhouse, kitchen, and servants' quarters in the basement; dining room, drawing room, master bedroom, and nursery on the ground floor; as well as a library and many spare bedrooms above. The house commanded a fine view of the city, thought Dawson; the air was good, and it promised to be "a delightful place in summer." Indeed, it was so much "in the country," that the family could find their rural pleasures there. (Eventually, however, the Dawsons escaped from the occasionally sweltering heat of Montreal's summers, to a home in northern Quebec at Little Métis.) Shortly thereafter, the grounds were covered by snow and frost. William was surprised by the warmth of their quarters, with thick stone walls and double windows that resisted the cold better than Pictou houses. Nonetheless, most necessities turned out to be costlier than in Nova Scotia; in Montreal, money bought only about half as much. Moreover, the only books generally available were cheap Scottish and American editions, reported Dawson to his bookselling father.[14] Indeed, William was soon urging his father to leave his home in Nova Scotia and join them in Montreal, which he was to do two years later.

MCGILL TRANSFORMED

It would be difficult to imagine an individual better prepared for the task at hand, better equipped for the challenges ahead. The youthful Dawson had already logged years of experience in the educational reform of Nova Scotia and New Brunswick. Soon he would be in a position to apply that knowledge to reshaping the educational institutions of Lower Canada. As well, his association with the University of Edinburgh, though brief, could provide useful lessons for his new career as principal and professor, particularly as McGill was steeped in Scottish, not English, educational tradition, where the university played an active (and non-denominational) role in daily life. Moreover, Dawson did not have to weigh the consequences of establishing residential colleges or appeasing sectarian interests; these decisions had already been dictated by the weight of experience and tradition.[15]

As the weeks passed in Montreal, William described his duties at McGill to his father in more detail. As professor of natural history, he joined five other professors who taught in the Faculty of Arts (medicine had eleven professors; law, four).[16] He lectured three times a week to a class of around fifty students (most of whom were studying medicine), whose number soon rose to sixty-five. Perhaps remembering the lectures of Robert Jameson, Dawson adopted the Scottish method of instruction. This meant the rejection of the Oxbridge tutorial system, and the expounding of his own views rather than a simple parroting of the textbook.[17] Apparently, the students were so keen about the course that they asked him to obtain copies of *Directions for Collecting Objects of Natural History*, a manual prepared by Spencer Baird at the Smithsonian Institution.[18] On alternate days, Dawson was temporarily required to teach a class in history, thereby replacing the mathematician originally drafted for the job.[19]

By the following January, William reported to his father that over a hundred students were enrolled in the college's three faculties, with the majority choosing medicine. The classes of the Faculty of Arts (as well as of the high school, where over 200 boys were taught) convened in the new building, Burnside Hall; the medical and law faculties met off-campus. Dawson and the college secretary conducted their business from the old building on campus.[20]

Although the Board of Governors had begun to recast McGill's image just before Dawson took over the principalship, the college assumed a new countenance under his direction. Physically, its ramshackle collection of deteriorating buildings began to take on the form of an institution of higher learning. Cognizant of the importance of keeping the grounds well manicured in order to attract the attention of the Montreal citizenry, Dawson decided to lay out gardens and plant trees at his own expense. In his view, an "active and hopeful" central campus was essential to obtaining the support of wealthy donors.[21] As Dawson states in his autobiography, the "embryonic condition" of the university meant that "everything depended on securing public support."[22]

The change in McGill's appearance during Dawson's first decade in Montreal owed much to his ability to tap the fortunes of the city's wealthy business elite, and then to channel this money to good educational purposes.[23] Indeed, this strategy was necessary because James McGill's original estate had been exhausted and appeals to the government for support ignored.[24] Dawson's Scottish background helped him gain access to the hearts, minds, and pockets of the predominantly Scottish "commercial aristocracy" who not only ruled Montreal but also controlled much of the wealth of Canada.[25] Using remarkable skill – indeed it was to become an art form – Dawson cultivated the tradition inaugurated by James McGill's original bequest.[26]

That Dawson and his predecessors were unable to obtain government assistance – either from the municipality, the province, or the Crown – materialized as something of a mixed blessing. The unfortunate aspect was that the university therefore pandered to an anglophone elite, largely cutting itself off from the support of the French-speaking majority. But as President Daniel Wilson reminded Dawson, he was lucky indeed to work with a "board of Trustees wholly independent of all political intrigue," unlike Wilson's own situation at the University of Toronto.[27]

Perhaps as an outsider unaware of the complex web of intermarriage and personal ambition that dominated the Montreal oligarchy, Dawson found the group's power somewhat less intimidating than others might have done, and more readily gained acceptance into their coteries. (Early in his tenure at McGill, he was to recollect "some slights" in Montreal society caused by the lowly position of the principalship, the Dawsons' commitment to evangelical religion, and their teetotalling practices.[28]) Certainly, he quickly became adept at exploiting the petty jealousies and competitiveness that both divided and drove this community. With each passing year, Dawson managed to elicit increasingly impressive levels of donations and bequests, something entirely unprecedented in the history of McGill. Few were content to remain anonymous, like graduate H.H. Wood, who sent an unsolicited cheque for $500.[29] More typical was the desire to exceed another's largesse: upon learning that George Drummond had given $4000, for example, a fellow Montreal businessman donated $5000.[30]

The Molson family emerged as a particularly important source of munificence during Dawson's early principalship (Mrs John Molson had been Dawson's fellow passenger in the steamer travelling back from Glasgow in 1855). In 1860, William Molson supplied a new wing to Burnside Hall (housing an assembly hall, library, and museum) and connected the three blocks, thus giving the campus "a harmonious suite of arts buildings." Perhaps even more important for the growing reputation of the college, the Molsons also initiated a series of benefactions that created professorships in a number of academic disciplines, beginning with English literature in 1857.[31] They supported, as well, Dawson's acquisition of natural history specimens, first for the small college collection and later for the magnificent new Peter Redpath Museum.

Sugar baron Peter Redpath, another major patron during Dawson's era at McGill, especially disliked the Molson family. Perhaps their rivalry fueled the spate of his donations. When John Henry Robinson Molson failed to provide enough financial support for the operating expenses of McGill's new natural history museum, Redpath explained that "obstinancy" was a family characteristic, and that Molson felt "his back has been rubbed the wrong way" by Redpath's endowment.[32] To Redpath, those who did not support philanthropic endeavours "have no more notion of such a duty [to help] than a man born blind has of colours."[33] But worse than a friend who gave too little was one whose contributions might outshine his own. Upon learning in England of a gift to help build a convocation hall and support a science school at McGill, Redpath anxiously wrote to Dawson for more details.[34]

Dawson shrewdly calculated the prize carrots to be dangled before potential benefactors. Without question, the juiciest brought a seat on the Board of Governors.[35] For example, after asking Montreal businessman Edward Greenshields to head a subscription drive for McGill graduates, Dawson held out the prospect of an appointment to the board, indicating that "any public action of this kind would tend in that direction."[36] Governorships, in particular, seem to have exacerbated jealousies and rivalries, all of which may have worked to the benefit of McGill's finances. Montreal Mayor William Workman, for instance, complained of his business partner David Torrance's appointment as governor. Unlike Torrance – a devotee of "some sect or persuasion, disliking all others" – Workman believed himself to be liberal, literary and cosmopolitan; moreover, he had contributed much to McGill and had not even objected to Peter Redpath's earlier nomination.[37] Dawson himself threatened to resign over the appointment of John Joseph Abbott, despite his impressive McGill pedigree and other powerful connections.[38]

Dawson arranged that the badly needed support would flow to students as well as faculty. Leading townspeople showed their appreciation of his labours by endowing medals, prizes, and bursaries.[39] But overshadowing even these generous contributions were the huge gifts given the university beginning in the 1880s – associated with the names of Redpath, Thomas Workman, and William C. Macdonald – which reached millions of dollars in the years leading up to Dawson's retirement. Around the same time, London barrister David Greenshield – so "full of benevolent spirit" – left $40,000, which was used to endow the principal's salary, over Dawson's modest protests.[40]

Despite these stimuli that would seem to have boosted the number of arts students, the medical and law faculties continued to outpace the arts. Nevertheless, as endowments increased, pride in the university as a whole grew markedly. A sense of solidarity emerged when the law and medical schools moved back into campus facilities during the 1860s and 1870s, and "the whole machinery of the institution was moving smoothly and regularly."[41] In his autobiography, Dawson was to reflect that the university "outlived, for the most part, its earlier trials and struggles" during this period. Nonetheless, none of these achievements, including the notable increases in students, faculty, and scholarships, could have been achieved without "considerable toil and many sacrifices" on Dawson's part.[42]

Without question, the university's financial affairs were conducted on a shoestring. Salary scales represented aspirations rather than realities, property had to be sold to raise revenue, and even a plan to introduce gas illumination had to be postponed because of its expense.[43] Professors were expected to work hard for their pay, although even Dawson complained of his own onerous teaching load of sixteen classes per week, which meant little time left over for research.[44] During the 1860s and 1870s, he devised varied ingenious strategies – including loaning money from his own pocket or forgoing a salary increase – to help weather successive financial crises.[45]

Dawson's support of McGill was never narrowly financial, however. His own large family, living as it did at the heart of the campus, played an active role in all extracurricular aspects of college affairs. William and Margaret treated students and faculty as part of their extended family. It was not unusual, for example, to see them nursing a sick student back to health.[46] And when Englishman Henry Bovey accepted the chair of engineering at McGill, the Dawsons offered him quarters in their home until he found lodging.[47] As for their soirées, to which all students in arts and science were invited, they were familiarly dubbed "Tea and Fossils," whether because of the aged specimens on display or the elderly women in attendance.[48] Indeed, the wife of one of McGill's governors praised the Dawsons for what they had done to elevate the character of Montreal society in general.[49]

When McGill's president and chancellor Charles D. Day first wrote to Dawson in 1855, he painted a bleak picture of the academic and physical resources of the university. He blamed McGill's problems on its attachment to "European-style classical education," a system not well adapted to Canadian life and, quite simply, "not the education the people want."[50] Dawson set out to change this orientation, and much of his work at McGill was dedicated to this end. In his inaugural address, he emphasized a pedagogical vision founded on Baconian utilitarianism – the triumph of modern science over antiquarian traditions. The pursuit of science led to a "double reward," according to Dawson, "first in the interest of its new facts and the ennobling general views to which it leads, and secondly, in its valuable and often unexpected applications."[51]

Rather than looking to Oxford, Cambridge, or even Edinburgh as models, Dawson expressed great admiration for Owens College, Manchester, with its "flourishing medical school and school of applied science," developed through the efforts of its citizenry.[52] He felt that McGill should import the ingredients of this success: the cultivation of the pure and applied sciences, which would win the support and open the purse-strings of forward-looking local merchants and entrepreneurs. Yet the recipe had to be adapted to the Canadian milieu, cautioned Dawson. Natural sciences like geology, meteorology, and botany would help to inventory untapped natural resources;[53] the development of mining and engineering would respond to the need to exploit and distribute these resources. Dawson sought to reshape McGill into a university with a professoriate and curriculum that used these issues to define academic priorities, albeit over the resistance of some of its governors.[54]

Although McGill's medical school had already established its reputation prior to Dawson's arrival in 1855, neither the natural nor physical sciences enjoyed any real presence in the arts faculty. The two professors who taught science, William Sutherland and Henry Aspinall Howe, were in fact "on loan" from other quarters (namely, the medical and high schools). As for James Barnston, who had been appointed professor of botany in 1857, he died before his first course of lectures was completed, and the position expired with him.[55] In effect, then, Dawson found himself responsible for the entire range of natural history sciences during his early years at McGill.

PHYSICAL SCIENCES WITH AN EMPHASIS ON METEOROLOGY

Soon Dawson was joined by two new professors in the physical sciences. Alexander Johnson, a graduate of Trinity College, Dublin, was appointed professor of mathematics and natural philosophy, a post he would hold for the next forty-six years. Charles Smallwood, M.D., resident of Ile Jésus – where he had established a small meteorological observatory – was named honorary professor of meterology, probably at the insistence of Dawson, who had come to know him well through the Natural History Society of Montreal.[56]

In 1863, Smallwood transferred his instruments to campus. E.T. Blackwell, president of the Grand Trunk Railway, built a stone tower (at a cost of around $2000) to house them, with the university providing the site. There Smallwood industriously conducted the McGill Observatory until his death a decade later. His most important duty was to determine the time astronomically; this became the standard for local businesses, the railroads, and a time ball in the harbour. McGill also telegraphed a time signal to Ottawa (for firing the noonday gun that regulated government time), and helped to synchronize the chronometers of ships docked in the port of Montreal.[57]

Upon Smallwood's death in 1873, his student and former assistant, Clement H. McLeod, was engaged to supervise the observatory. With Smallwood's demise, George Templeman Kingston of the Toronto

Observatory put forth criticisms of his work. (This may explain why McGill's time service ceased operation for a while, becoming the purview of a private astronomical observatory in Montreal belonging to businessman and amateur astronomer Charles Blackman.[58]) Kingston complained of numerous and frequent errors in arithmetic, saying that he "felt ashamed of Canada when I forwarded the Montreal telegrams to Washington." Compounding Kingston's opinion that Smallwood had proved to be "utterly disqualified for scientific work" was McGill Observatory's poor location, too close to the city's smoke and lights. Kingston even offered to loan equipment and to send a mechanic (free of charge) to calibrate McGill's instruments.[59]

Under McLeod's direction, the state of the McGill Observatory began to improve. A bequest of astronomical instruments (a large equatorial telescope, a transit instrument, and an astronomical clock) came from Blackman, who was closing up his observatory and moving to the United States. The McGill Observatory was partially rebuilt in order to accommodate the new telescope. As a result of these improvements, the time signal was restored, thus permitting McGill to collect payments again for time determination from the Montreal harbour commissioners and the railways.[60] The university gained further prestige by participating in the observation of the transit of Venus in 1882; as Dawson explained, McGill would receive credit for observations made with its instruments (although, in the event, bad weather thwarted their work).[61] Unlike Smallwood's records, McLeod's observations became renowned for their accuracy, and his work over the decades helped to establish a good reputation for the McGill Observatory. Nonetheless (and despite the rhetorical assurances of Dawson on several occasions), McGill never emerged as a centre of research or teaching in astronomy.[62]

Unfortunately, the fate of natural philosophy was generally undistinguished during most of Johnson's tenure and Dawson's principalship. This all changed in 1890, however, when Macdonald donated the funds to build a magnificent new physics building, as well as $50,000 to endow a chair in experimental physics. A new era of world-class accomplishment in the physical sciences was about to begin just as Dawson stepped down as principal in 1893.[63]

THE DEPARTMENT OF APPLIED SCIENCES

If, under Dawson, education at McGill was to be made responsive to the needs of an emerging Canadian nation, part of his mission was to establish the applied sciences on a secure footing. Agriculture and mining were obviously relevant to the exploitation of Canadian terrain, while the railroads (and the expertise in engineering required to build and maintain them) permitted transportation of extracted goods across the vast continent. Indeed, thought Dawson, a concentration on technology rather than science might win over such potential donors as Hugh Allen, the steamship magnate, who had been skeptical about the general utility of McGill.[64]

Dawson himself attached great importance to teaching agriculture. From the start, in addition to his many other responsibilities, he asked to be appointed professor of agriculture (without pay), in order to offer an optional course to advanced students.[65] But the subject never really developed into a viable academic offering during Dawson's principalship. Dawson attributed the low enrolments in the agriculture courses – patterned after those at Edinburgh and Yale universities – to "the absence of scholarships, prizes, and other inducements." (Indeed, the Canadian entrepreneurs who patronized McGill at that time were preoccupied with extractive industries rather than farming.) A serious commitment to the field would have required the establishment of an agricultural library and a botanical garden as well.[66]

The beginnings of engineering seemed, at first, more auspicious. In 1857, Mark Hamilton was named professor of civil engineering, as well as professor of road and railroad engineering. The intention was to separate the appointments into two distinct chairs as demand grew. But after a brief period of gradual growth in enrolment, enthusiasm waned (largely as a result of a general economic decline), and Hamilton's appointment was terminated in 1863.[67]

Despite this dismal record, Dawson was convinced that "a thoroughly equipped school of mining, metallurgy and assaying, engineering and land surveying" was necessary at McGill. He constantly petitioned government officials at both federal and provincial levels to provide grants to this end.[68] Nonetheless, he privately wished that the university had the means "to do the *whole* ourselves," thereby reaping all the credit for McGill from this stimulation of the mining industry "altogether out of proportion to its cost."[69] Alarmed at the frequency and stridency of Dawson's requests, governor and lawyer Christopher Dunkin counselled him "that with a little patience we shall be able to get the substance of what we want."[70]

Ultimately, Dawson's wishes were fulfilled. The Department of Applied Science was resurrected not by governmental support, but through personal subscriptions of the McGill Board of Governors, totalling nearly $7500.[71] Dawson came to realize that Yale's Sheffield Scientific School, Owens College in Manchester, and the Royal College of Chemistry in London all revealed the same pattern: enlightened benefactors "taking the first step."[72] In addition, an endowment by

J.H.R. Molson, Peter Redpath, and George H. Frothingham permitted Bernard James Harrington, a McGill graduate and Yale Ph.D. (and Dawson's future son-in-law), to be hired as lecturer in assaying and mining in the Department of Practical Science. Harrington was delighted to return to Canada, anticipating if not a brilliant career, at least one where he could work "perseveringly and faithfully."[73] (Only Dawson's son, George Mercer Dawson, who had attended mining schools in London and Paris, seemed piqued by Harrington's appointment, wondering how he could teach mining when he had "never been down one [a mine] in his life."[74]) Shortly thereafter, George Frederick Armstrong, a Cambridge graduate, was named professor of civil engineering. Soon the departmental name changed to applied science, and its status raised to that of a faculty. Gilbert Prout Girdwood, M.D. (graduate of University College, London) was asked to teach practical chemistry (as he already did in the medical faculty) and Archibald Duff (B.A., McGill), mathematics. When Armstrong resigned in 1877, he was replaced by another Cambridge graduate, Henry T. Bovey.[75]

By the 1870s, unlike the situation a little more than a decade earlier, student enrolment in the arts faculty began to increase rapidly. Although Dawson worried about the effect of the Geological Survey of Canada's move to Ottawa, the constant trickle of donations from Montreal businessmen seemed to eliminate any serious financial obstacles.[76] Nonetheless, even the self-effacing Harrington found reason to complain of the governors' parsimony in purchasing chemical apparatus, which Harrington believed should be placed "in the same category as coal and other necessaries which must be obtained every year if the college is to be kept alive."[77] By 1882, the faculty boasted ten professors and forty-four students.[78]

The fortunes of applied science reached a new era of security in 1889, when Thomas Workman bequeathed a building and a professorship in mechanical engineering. But even this large donation was eclipsed when, in the early 1890s, Macdonald underwrote a new engineering building and a chair of electrical engineering. On the eve of Dawson's retirement as principal, a bright new future in applied science at McGill was on the horizon.[79]

THE DEVELOPMENT OF NATURAL HISTORY SCIENCES AT MCGILL

Given Dawson's unflagging interest in all aspects of natural history sciences, it is hardly surprising that geology, biology, and paleontology achieved special prominence during his principalship. Unlike other academic environments, however, such as the University of

Edinburgh where natural history served as a handmaiden in the medical faculty, at McGill these professorships and courses belonged to the arts faculty. This arrangement was doubly beneficial to the arts, because the fees collected from medical students could be invested outside the medical faculty – that is, within arts or wherever Dawson saw fit. In the tradition of Jameson at Edinburgh, Dawson emphasized the geological component of his chair, and soon sought to establish the discipline on a stronger footing within the university.

In the early 1870s, just as Montreal's business community was experiencing a period of economic depression that would discourage even the most zealous fundraiser, Dawson called for a public meeting of one hundred gentlemen "to devise means for giving a new impulse to the college." He asked William Logan, director of the Geological Survey of Canada, to spearhead a campaign to endow a chair in geology for $20,000. Of course, the professorship would carry Logan's name; it might even, said Dawson, attract first-rate scientific talent to McGill in the near future.[80] Logan consented (he raised $1000 from his brother James), requiring only that Dawson himself be the first incumbent, and that he (Logan) be consulted about a successor if he should outlive Dawson.[81] By 1883, thanks to the proceeds of the endowment, Dawson was receiving a salary of $2800 as the Logan professor.[82]

Believing that his own professorship comprised paleontology as well as geology (and that zoology was being adequately treated in the medical faculty), Dawson identified botany as the next chair for endowment. To help establish the professorship, Dawson wrote the amateur naturalist Charles Gibb (who lived in the Eastern Townships), asking him to mobilize "friends of the university" to secure an additional thousand dollars on top of the thousand that had already been pledged.[83] Gibb, although not "sanguine of success," offered Dawson a hundred dollars from his own pocket.[84]

Dawson sought, as well, to attract to the position one of Asa Gray's star students, the American botanist David Penhallow, who was then working at an agricultural experiment station in New York state after returning from teaching at the Imperial College of Agriculture in Japan. But to do so, McGill needed a permanent chair. Gray's recommendation of Penhallow in fact led him to mend fences with Dawson (a dispute over scientific matters had earlier strained their relationship).[85] Dawson also solicited candidates from elsewhere in Canada, although Geological Survey botanist John Macoun, explaining that botanical knowledge in Ontario was "below par," refused to recommend anyone,[86] and the qualifications of two suggested New Brunswick naturalists – James Vroom and George Hay (brother-in-law of the late Charles F. Hartt, geologist and professor at Columbia University) – placed them far below Penhallow.[87]

In any event, the prestige of Penhallow's qualifications attracted sufficient donations to make a professorship a possibility, and, in turn, this eventuality attracted him to Montreal. Bernard Harrington and his wife, Dawson's daughter Anna, sent Dawson (who was in Europe when Penhallow assumed his duties) glowing reports about the new professor (except for one incident that found him besieged by medical students clutching frying pans after he had interrupted a snowball fight). Anna warned, though, that if Penhallow were not apprised of a more permanent arrangement, he might "just slip through your fingers which would be a pity."[88] In Penhallow's description to Dawson of how he conducted his courses, he depicted his medical students as learning about the connections between botany and anatomy, and his arts students as concentrating on economic and industrial applications.[89] Penhallow's utilitarian approach to diffusing botanical knowledge coincided perfectly with that of McGill's absent principal.

MCGILL'S NATURAL HISTORY COLLECTIONS

The opportunity to curate the botanical materials brought together in the Peter Redpath Museum, which had just opened its doors a year before his arrival in Montreal, was a major factor in Penhallow's decision to remain at McGill. Most of the collections in the magnificent new museum building had evolved from the individual objects assembled by Dawson years earlier. Indeed, when he first arrived in Montreal, Dawson had vowed to Spencer Baird at the Smithsonian Institution that he would create a respectable museum on the McGill campus.[90] The pledge was especially ambitious because, at the time, the university possessed but a single fossil.[91]

Dawson managed to have a room set aside to house a modest museum; he expanded the collections by means of occasional gifts of money, the fees collected from his lectures to medical students, and the museum fund established by banker and brewer William Molson. Specimens came from Montreal residents, such as Philip P. Carpenter, who gave an extensive collection of shells valued at $7000, and the medical professor Andrew Fernando Holmes, who provided a herbarium.[92] As well, Dawson gathered fossils and rocks during the summer holidays, some of which went directly to the museum, while others were used for exchanges with such major establishments as Kew Gardens and the Smithsonian Institution.[93]

By 1862, Dawson boasted that McGill's museum held 10,000 natural history specimens, arranged to illustrate successive lecture topics in that subject. In addition to its pedagogical function, the collection was available to local amateur naturalists. Dawson was always ready to promote the merits of the museum in order to loosen the pursestrings of a potential donor. He delicately suggested to Anne Molson, for example, that if her father William was considering any additional benefaction to McGill, the needs of the museum "might give direction to anything he may design."[94] Eventually, Dawson's zealous activity led to an overcrowding of the natural history collections, requiring him to put much in storage and to keep the "working collections" at his own residence.[95] Nonetheless, McGill did not intend to amass a "large general collection," rivaling those of the two other Montreal-based institutions: the Geological Survey of Canada and the Natural History Society. Dawson wished to expand McGill's holdings – strong in Paleozoic fossils, but weak in American Mesozoic and Tertiary fossils – only in areas not already represented in the collections of these other two museums.[96]

The presence of three natural history museums placed Montreal far ahead of other Canadian cities in terms of scientific resources. Dawson believed that such an "accumulation of museums" always preceded the establishment of a "higher grade of scientific schools." Training in science, unlike other branches of knowledge, required "large preparatory appliances in collections and apparatus," not just books and teachers, said Dawson, perhaps remembering the fine specimens at Pictou Academy.[97] For these reasons, Dawson attached great importance to the continuing existence and expansion of Montreal's museums, particularly as they contributed to the growth and scientific importance of McGill.

PREPARATIONS FOR THE PETER REDPATH MUSEUM

By the late 1870s, Dawson's view of the purpose of McGill's natural history museum and its relationship to others in Canada had changed dramatically. Foremost in precipitating this change was the Dominion government's decision to transfer the Geological Survey of Canada and its museum to Ottawa (the new national capital, widely perceived as a provincial backwater). Dawson's fury over this development fueled opposition to it in Montreal, and feelings smoldered from the first announcement of the plan in 1877 until the actual move during April and May of 1881.[98] Indeed, he termed the move an "act of gross vandalism."

Dawson encountered remarkable indifference to the move among some of his former friends and associates on the survey staff. His conception of the scope and function of McGill's collections shifted accordingly. No longer was he content to build a modest museum; instead he aimed to establish "a better collection illustrative of Canadian geology" than that of the Geological Survey, and in less than a year. What gave conviction to Dawson's determination was Peter Redpath's offer to provide McGill with a museum building that would be "the best of its kind in Canada."[99]

On first hearing of Redpath's offer, Dawson thought of the needs of the applied science students and faculty, working away without a "decent building." Apparently, Redpath disagreed with Dawson's suggestion to make the museum project contingent on the funding of a science building by another benefactor, or to transmute his offer of a museum into a donation for a science facility. Redpath seems to have wanted to respond more to Dawson's personal needs than to the institution's, to commemorate Dawson's twenty-five-year term as principal with something more than "a superb silver watch with your name elegantly engraved upon it." More importantly, Redpath also hoped to dissuade Dawson from accepting a post at Princeton University.[100] With his gift of a museum costing around $140,000, Redpath initiated a new level of private bequests to the university[101] and a new era in funding.

By late summer 1880, the new museum building was underway. Peter Redpath himself supervised the laying of its cut-stone foundation.[102] A year later, the roof was in place.[103] As work on the museum's edifice progressed, Dawson laboured zealously to build up the collections. His own cabinet of nearly 10,000 Canadian rocks and fossils (valued at $5000) formed the nucleus. The heirs of William Logan, as part of a complicated manoeuvre related to the transfer of the survey, donated $4500 for a collection in his memory. With these funds, Logan's former assistant, James Richardson, was hired to collect duplicates of Canadian specimens held exclusively by the Geological Survey museum.[104] J.H.R. Molson gave $500 a year to purchase collections on the auction block.[105]

A key player in Dawson's museum building was his son, an employee of the Geological Survey of Canada. During the summer of 1882, George Mercer Dawson toured Europe, sent home timely information about continental museums, and cultivated useful contacts abroad. Particularly impressed by the provincial museums of France, George urged his father to add "a small typical local collection" to the Montreal museum, "with [a] map to accompany it so that anyone could go to the precise spot at which points of importance exist." In

Bonn, he visited the geological "merchant" August Krantz, whose immense collection furnished specimens at much less expense than those from London. Moreover, George found that he could procure rocks and minerals in the French countryside for as little as twenty-five centimes apiece.[106].

Dawson acquired other natural history objects by exchanging with institutions and individuals across Canada, the United States, and Europe. Cordial relations were soon restored with the Geological Survey of Canada: three months after the move to Ottawa, Dawson was receiving shipments from them; and within the next year, the Redpath Museum was reciprocating. As well, Dawson established exchange ententes with keepers at leading museums abroad, such as Henry Woodward, head of the British Museum's Department of Geology. But it was especially to "surveys and private collectors in the United States" that Dawson looked.[107]

Earlier Dawson had warned Member of Parliament Thomas White that the result of the Dominion government's transfer of the Geological Survey of Canada museum was "to annex us practically to the United States." Perhaps Dawson meant that he himself would look south for support. He eagerly swapped fossils with the foremost amateurs in the United States, including R.D. Lacoe; with state museum directors, such as James Hall in New York; and with the most distinguished local societies, such as the Boston Society of Natural History through the efforts of its officers, Alpheus Hyatt and Samuel Scudder. He also arranged exchanges with curators in the largest museums in the land, namely, Richard Rathbun at the National Museum in Washington and R.P. Whitfield at the American Museum of Natural History in New York. Assistant Secretary Spencer Baird enticed Dawson to aid the Smithsonian Institution's expedition to Ungava Bay by promising him their first series of duplicates – better than the specimens going to Ottawa – for the Redpath Museum.[108]

THE REDPATH'S OPENING DAY, 1882

Peter Redpath chided Dawson that his insatiable appetite for "stones and bones and skeletons of all kinds" might overwhelm the capacity of the new building. As the August 1882 opening date approached, preparations reached a feverish pitch. Dawson and his son-in-law Bernard Harrington, then professor of chemistry and mineralogy, sacrificed their summer holidays to label and arrange specimens in their cases. To speed the work of assistant curator Thomas Curry, piano-factory employee Paul Kuetzing was hired to mount and renovate vertebrate animals. Edwin Howell, Henry Ward's partner in the taxidermy firm located in Rochester, New York, travelled to Montreal to set up a copy of the British Museum Megatherium (a status symbol for new museums) and other large objects. And a number of McGill students and graduates volunteered to help transfer the university's collections into the new museum building.[109]

Dawson called the sight that greeted the 2000 guests at the formal reception inaugurating the building "the greatest gift ever made by a Canadian to the cause of natural science, and ... the noblest building dedicated to that end in the Dominion." Journalists acclaimed the museum as "one of the greatest architectural beauties of Montreal," lauding "its simplicity of outline and marked constructive strength."[110] The Grecian-style exterior, built of limestone quarried near Montreal, represented conventional architectural practice. (By this time, the new natural history museums in London and Paris had turned away from classical traditions, and had incorporated biological symbolism into their Gothic or Romanesque façades.) Nor were the dimensions of the building (132 feet long and 72 feet wide) remarkable by world standards – the American Museum of Natural History covered thirteen acres of floor space. Still, Redpath Museum exhibited pleasing external proportions and a well-designed interior plan, with adequate space for an instructional series of natural history specimens.[111]

Entering the Peter Redpath Museum, the visitor could see at the back of the ground floor a handsome lecture theatre with seats for 200 students. Rooms closer to the front of the building would soon accommodate a herbarium, reference library, classroom, boardroom, and office. At the right side of the entrance, a staircase fitted out with archaeological objects and large slabs of fossil footprints on the landing led to the main floor or "Great Museum Hall." Ward's imposing cast of the *Megatherium* distinguished this main floor, with displays of paleontological, mineralogical, and geological specimens. Fossils along the centre and at either side were arranged according to their progression in geological time, their botanical or zoological classification subordinate to this organization. Thus the visitor could view the general order of geological succession or trace any group of animals or plants through several formations. The second floor of the museum – the gallery of the great hall – contained zoological material, both representative types and local forms. Invertebrates were stored in table cases, while vertebrates were displayed in upright cases. The basement contained a laboratory where specimens could be prepared and stored.[112] Dawson, describing the museum to Alexander Agassiz at the Museum of Comparative Zoology, called it "the first properly organized thing of the kind [that] we have had in Canada."[113]

THE REDPATH'S FIRST DECADE

A small but distinguished committee chaired by Dawson managed the affairs of the Peter Redpath Museum. In addition to McGill's other natural history professors (only Harrington at first, but joined by Penhallow in 1883), three members of the corporation sat on the committee. The Board of Governors elected Peter Redpath to the group in January 1882. Given Redpath's anticipated long absence in England, the committee added J.H.R. Molson to their number. Unlike Redpath, Molson took the responsibility seriously, attending the bimonthly meetings regularly and reporting back to the other members on various matters. For the next five years the composition of the group remained fixed, once Professor John Clark Murray of moral philosophy had replaced the deceased dean of the medical faculty, George W. Campbell, in August 1882. By the late 1880s, however, resignations and deaths had altered the committee, which by then was meeting at quarterly intervals.[114]

The Redpath Museum Committee worked with remarkably scanty financial resources. A small portion of medical students' fees (several hundred dollars per year) came from the university in exchange for the students' use of the building's laboratory facilities. On occasion the Board of Governors advanced funds so the museum could balance its accounts; these amounts had to be repaid, however (each operation of the university supposedly being financially self-supporting). Because McGill had agreed to preserve the museum, according to the terms of Peter Redpath's donation, the corporation was to pay for repairs and improvements. The museum, for its part, was held responsible for general maintenance. A somewhat arbitrary and bizarre division of responsibility ensued: for example, McGill paid for snow removal from the roof; the museum, from the grounds. The university took charge of painting the roof, while the museum oversaw the varnishing of woodwork around the windows. As well, the museum received revenue from the 25-cent admission charge, levied on all visitors except university staff, McGill graduates, schoolteachers, and clergymen. Money also accrued from interest on the various museum funds and from fees paid by the Ladies' Educational Association (about $100 per year) for lectures delivered in the museum theatre.[115]

Perhaps because of their firsthand knowledge of the museum's dire financial situation – it seldom moved outside the red – members of the museum committee gave generously of their money as well as their time. In addition to Redpath's annual grant of $1000 for maintenance of the museum building, Louisa Molson contributed $2000 to establish a fund to pay for the salary of Thomas Curry, the assistant curator. Louisa's husband, J.H.R. Molson, donated at least $500, and sometimes as much as $1000, a year for the purchase of otherwise unobtainable collections.

Molson's generosity enabled Dawson to buy rocks and fossils from naturalists and dealers overseas, including Anton Dohrn at the Zoological Station in Naples, the elderly Edward Charlesworth in London, Charles Moore in Liverpool, and August Krantz in Bonn. By the late 1880s, the museum collections were valued at nearly $60,000.[116] Generally, however, Dawson relied on donations from friends in Montreal, elsewhere in Canada, and abroad, all of which were duly acknowledged in the annual reports of the museum and at quarterly intervals in the Montreal *Gazette*. Some of these acquisitions, such as Lieutenant-Colonel Charles Coote Grant's collection of Silurian fossils, were valuable additions to the museum's inventory. Yet such items as stuffed song sparrows and Baltimore Orioles – accepted so that potential patrons would not be discouraged or offended – strained the already limited museum resources of display and preservation.

The salaries and fees of those who cared for these materials and maintained their surroundings accounted for a major source of expenditure. In addition to Curry (who mounted, labelled, arranged, catalogued, and occasionally collected specimens), Edward Ardley became a permanent employee of the museum. As caretaker, he had initially earned only $30 a month plus lodging in the museum basement (including fuel gas); over the years, however, his tasks became increasingly skilled and specialized. The museum committee purchased him a set of carpenter's tools in 1886 for his construction of display stands and shelves. Three years later, because of the increasing size of the collections and Curry's failing health, Ardley was regularly cleaning and mounting specimens. He also learned to operate a lathe in order to slice sections of rocks and fossils.

Other hands were hired on a casual basis to carry out specific assignments. Several McGill graduates arranged, labelled, and catalogued collections of insects and fossils. Although Henry Ward set up a gorilla skeleton acquired from Liverpool, Jules F.D. Bailly usually acted as resident taxidermist. To him fell the honour of mounting the skeleton of the bison shot by Molson and Dawson between Calgary and Medicine Hat.[117] Another Montrealer, George Roberts, built the display cases increasingly required as the collections grew. When the museum lacked the $600 to purchase cases for certain botanical specimens, Roberts' offer to defer payment was eagerly accepted. Unfortunately for the poor carpenter, a year elapsed before he received even half the amount owed.[118]

Although the primary function of the Peter Redpath Museum was to serve McGill students and faculty, many educational and professional organizations also enjoyed its facilities. In the early 1880s, the American and British associations for the advancement of science held geological sessions in the lecture theatre and receptions in the Great Hall during their Montreal meetings. The Protestant Association of Teachers and the Canadian Society of Civil Engineers met in the Redpath Museum as well. There, too, the Ladies' Educational Association heard lectures on botany, zoology, and the "geology of Bible Lands." By the early 1890s, however, evening entertainments in the museum had ceased because of the great risk of fire. Taking into account its uses by the university and the average annual attendance of around 2000 during these years, the Redpath Museum could claim to be "the foremost educational institution in Canada."[119] It was Peter Redpath's final encomium, however, that indicates how well Dawson fulfilled the vision of its benefactor: "You have made the Museum useful beyond my most sanguine expectations and I cannot but be gratified by the feeling that its cost was not ill bestowed."[120]

If for nothing else, Dawson should be celebrated for his revitalization of McGill University. He transformed it into a modern institution of higher education, and made it responsive to the needs of a burgeoning colonial city. As the *Times* of London summed up Dawson's achievements at McGill, "The scientific side of the University (if we except the Medical Faculty) may be described as Sir William Dawson's creation."[121] Numerous attestations by important correspondents (as well as by every one of McGill's historians) indicate that there was widespread recognition of Dawson's remarkable accomplishments years before his death. McGill's chancellor Charles Dewey Day reflected at one point that "the university which has caused both of us so much anxiety and you so much labour is reaching upward and justifies the hope of brilliant progress not very far off."[122] Civil servant James M. Lemoine accurately predicted (just before Dawson's knighthood) that he might be in line for some "royal recognition" for his striking success at McGill.[123] But Dawson's remarkable stewardship of the university is perhaps best summed up by his old patron Sir Edmund Head, just five years after his arrival in Montreal: "I believe that among the best things which I was able to do in Canada, I ought to reckon my taking advantage of the opportunity of placing you in the position which you now occupy."[124]

6 "Stand by and grumble"[1]

However successfully Dawson promoted a vision of McGill University where academic excellence was grounded upon non-sectarian cooperation, he was faced with a fundamental paradox: the university functioned as an enclave of English Protestant higher learning, but was situated within the milieu of Roman Catholic, French-speaking Quebec. This fact of life did not usually hinder Dawson in promoting the expansion of McGill, since its potential isolation from provincial and municipal mainstreams proved a drawing card for the financial support of Scots Presbyterian entrepreneurs such as Peter Redpath. Yet whenever Dawson wanted or needed to broaden his network of funding, this paradox became a liability. French Catholic civil servants and politicians, not surprisingly, failed to be overwhelmingly sympathetic to McGill's plight. Particularly as Dawson became more involved with Protestant education at the secondary level, he was forced to confront this anomalous position on a regular basis.

In 1867, with Confederation, the precarious state of Protestant education in Lower Canada was about to be fundamentally altered. Even on the eve of Canadian unification it remained unclear how the English Protestant minority and their needs would be accommodated within the new political and legal structures. In Dawson's view, education was an issue of fundamental concern in a country such as Canada, given its lack of national heroes such as George Washington and William Tell, as well as the absence of "stirring memories of great deeds done or sufferings endured." To Dawson, these myths and legends acted to bring together "discordant interests, races, and creeds."

Without them, he believed, Canada would tend to respond to the decentralizing tendencies of petty and parochial forces. Dawson opined that strength and unity could not come from "popular indifference on the one hand, and mere declamation on the other." Only "enlightened public opinion" could fill the void; this in turn depended on "the cultivation of minds fitted to guide aright the destinies of the country, and to reconcile its jarring interests without any fatal sacrifice of truth and right."[2] All rhetoric aside, Dawson could not have been more "deeply interested" in these questions.[3] Not only was the future of his country at stake, but so was everything he had worked to build in Montreal over the past decade.

PROTESTANT EDUCATION IN QUEBEC

Responding to a unique set of political and economic circumstances (which he recounts in his autobiography), Dawson had helped to establish a teachers' college shortly after his arrival in Montreal: the McGill Normal School. This school served as a link between the university and "the higher schools of the English and Protestant population," thus giving their branch of the school system "unity and strength." Although there could be no doubt of the importance of the McGill Normal School for the cause of Protestant education in Quebec, Dawson regretted the fact that he was forced by subsequent events to assume the principalship (and a professorship) at the school for the next twelve years. He calculated that the two months a year he lost to McGill Normal School matters cost his original research dearly "during several of the best years of my life."[4]

Dawson also devoted his time and energy to ameliorating the affairs of the provincial department of education. Through this department he aimed to provide "a uniform course of study leading to the University matriculation," and also to stimulate the establishment and improvement of English schools at the primary and secondary levels.[5] Indeed, this department (the Council of Public Instruction) had just been established in 1856, responding to the need to impose standards on Lower Canada's decentralized and fragmented (by language and faith) public educational system.[6] Dawson committed himself firmly to the goal of a comprehensive educational system, the importance of which he realized from his work in Nova Scotia and New Brunswick. Although the dual confessional and linguistic complexion of Quebec made this ideal more difficult to achieve, Dawson nevertheless sought a unity of purpose and a plan among the English Protestant schools at all levels.[7] These schools were represented by a Protestant board, with a secretary; its analogue, the Roman Catholic board, gave a seat to each of the Quebec bishops. Both boards fell under the jurisdiction of Quebec's superintendent of education.[8]

Once members of the English-speaking community of Montreal recognized the inevitability of Confederation, they realized that they were about to become "a permanent linguistic and religious minority" in Quebec. The provincial boundary now separated them by law from their coreligionists in Ontario, who were there a majority.[9] For his part, Dawson preferred a "federation of the colonies with Great Britain" over confederation, which he saw as "destructive to all good in these colonies."[10] Dawson painted a dramatic picture to the geologist William Logan: they were about to pass through a "political revolution," in which they would be threatened with the possibility of the colleges and schools of Lower Canada being transferred to the French-speaking majority. He warned that circumstances have given the French "every excuse to crush us, or at least to place us under greater disadvantages than at present," while the English can only "stand by and grumble" and hope for some "guarantees."[11]

Dawson even asked his old geological mentor Charles Lyell to press the concerns of Quebec Protestants upon anyone of influence. He worried that their educational interests were about to be delivered into the hands of the Jesuits "who now rule the French here with a rod of iron, and through them oppress the English population." He wondered whether Quebec might not be reduced to the condition of Spain or the papal states.[12] Nor did the anti-evolutionist Dawson hesitate to employ Darwinian imagery in this instance. He claimed that the English Protestant educational institutions faced extinction, and that he was being forced to become engaged in a "struggle for existence."[13] He wrote that he was "despondent," that he could only "trust that divine providence will work good out of evil for us."[14]

These occasions of hysterical prose appear to belie Dawson's ultimate resignation to the inevitable. But he was to remain deeply disturbed at, and confounded by, the "unaccountable apathy" of the Protestant community in Quebec. In the face of this apathy, he threatened to "abandon" his activities in the area of "general education" and devote his attention to science.[15] He still could not accept the analogy drawn by some Protestants between their minority interests in Lower Canada and those of the Roman Catholics in Upper Canada. In Dawson's view and experience, Quebec's English-speakers could expect no support from other minority groups in Canada. He had little hope, moreover, that the new federal government would rise to their defence in educational matters.[16] He believed, in essence, that there remained no alternative to the active promotion of the cause of English Protestant education within the boundaries of

Quebec province and to the forging of alliances with anyone who shared the same convictions.

Dawson came to work closely with the spokesman for Protestant interests in Quebec, Henry Hopper Miles, who had left a professorship in mathematics and natural philosophy at Bishop's College in Lennoxville (Quebec) to serve as acting secretary of the Council of Public Instruction and secretary of its Protestant Committee. Miles alerted Dawson, for example, when a "senseless and some would say unjust mode of dividing the aid for higher education" reduced the Protestant grant by half, redistributing these funds to the Catholics. Miles considered resigning over the incident, for he believed that the Protestants would get nothing more than what they could "extort."[17]

Another ally was the Montreal rabbi and McGill professor Abraham De Sola. De Sola, too, complained that the educational appropriations were "so shamefully partial and unfair," and objected that the taxes from the "considerable" amount of Jewish property had been taken from the Protestants and directed into the general funds. De Sola advised Dawson, "that like every other act of high handed tyranny, oppression and injustice it will certainly ultimately recoil on its perpetrators."[18] In the end, in part due to the surprising support of Montreal's Catholic school board, the law was amended to give the Protestant board all the monies levied on Protestant property in the city. In Montreal, the taxes would not be apportioned to the boards on the basis of population, as occurred elsewhere in Quebec.[19]

In 1876, Dawson was appointed to the Protestant section of the Council of Public Instruction.[20] (By this time, the Protestant committee, like the Catholic, operated autonomously.[21]) The job required "great zeal and energy," for the committee could expect to meet "constant passive obstruction," according to fellow member, Charles Day.[22] Presumably, Dawson's appointment was to prevent the "sacrifice" of Protestant interests and give them "a good measure." Others saw the committee as "guided almost entirely by your [Dawson's] views on any given subject."[23] Indeed, every appointment was seen as a potentially controversial and contentious issue in this politically charged atmosphere. For instance, the nomination of George Weir was decried (or applauded, depending on one's perspective), for he had long been considered an "enemy" by the Catholics.[24] The dismissal of Presbyterian clergyman Donald MacVicar was interpreted by Dawson as a "heavy blow" to the "struggling cause of Protestant education," although he was assured that MacVicar's term had simply expired.[25]

Dawson's regular attendance at committee meetings has led one historian to commend him for assiduousness.[26] Nonetheless, he and his colleagues believed that, in these difficult circumstances, individual personalities could only go so far or do so much. One member complained that until the money voted for Protestant education was "put under its control and at its disposal," the committee wielded no real power.[27] Another maintained that their problems were simply endemic to governments everywhere, since "the interests of the higher civilization go [to the wall] before the bristling noisy aggressive demands of material progress."[28]

Before he had logged a decade on the Protestant committee, Dawson resigned his seat. As he had earlier complained to Lyell, he still believed that "a thoroughly despicable and I fear priest-ridden economy prevails in all educational matters in the government of the country.[29] His friends resolved that Dawson's departure, "removes from our midst a great champion of the rights of the Protestant part of the population of this Province," and that Dawson's labours in the cause of higher education "have left a mark which can never be effaced."[30]

Perhaps Dawson simply tired of being the lone voice in the wilderness, and of losing so much time that might have been spent on his other responsibilities and interests. Richard W. Heneker, chancellor of Bishop's College and associate member of the Protestant Committee, asked Dawson: "Is there no high minded public spirited lawyer in Montreal, of sufficiently high standing to command attention – and at the same time to interest himself in public Education?"[31] But no one came to Dawson's assistance, despite Prime Minister John A. Macdonald's assurance that "by persistent complaint and remonstrance from the Protestant inhabitants of your Province you will do much ..."[32] Dawson was even consulted about how the setting up of Protestant schools in Quebec might affect the establishment of Catholic schools in Nova Scotia, as opposed to the provision of a nonsecular public system open to all.[33] Although Dawson's sense of mission and purpose called on him to protect the rights of English Protestants, Quebec reality so interfered with his vision that he could not help but become discouraged and despondent.

THE REMOVAL OF THE MUSEUM OF THE GEOLOGICAL SURVEY OF CANADA

If the English minority in Quebec felt betrayed by the French majority in the operation of the provincial educational system, they were equally unhappy about their treatment at the hands of the new Dominion government. Earlier, they had taken for granted the continued existence of federal institutions in Montreal, a vestige of the days when Montreal alternated with Toronto as the capital city. Events would prove this complacence to be ill founded. As early as

1865 (in the same breath as he was reflecting on the fate of English education in Quebec), Dawson asked Logan about the place of the Geological Survey of Canada (GSC) under the new constitution: would he agree to have it "split up between the local governments," or did he want it placed under the general government?[34] Charles Day connected the two subjects as well, commiserating that whether the dispute centred on Protestant education or the GSC Museum, the French members of the government would do nothing to help the beleaguered English in Quebec.[35]

The issue of the GSC preoccupied Dawson because he had always enjoyed a close, paternalistic relationship with the group. On numerous occasions, he had been asked to recommend potential employees to the director. A word from Dawson usually secured a berth, although he was reminded by one politician "that the imperative mood is apt to give offense" to government officials.[36] (His own son, George Mercer Dawson, joined the GSC staff in 1875, and worked his way up through the ranks to become director in 1895.) Logan and his successor, Alfred Richard Cecil Selwyn, never hesitated to place the "keys of the [museum] cases" at Dawson's disposal, nor to consult him about acquiring particular collections.[37] When Dawson applied to the Royal Society of London for an exceptional publication grant, Lyell suggested that mention of how much he had done for the GSC under Logan's directorship would likely win him support.[38] In the end, however, Dawson turned to the GSC itself to publish his long monographs on Devonian plants and other topics.[39]

The possibility of transferring the GSC, and especially its museum, to Ottawa was raised as early as 1867, the year of Confederation. McGill authorities believed that they had assembled a critical mass of facilities – especially the new Department of Applied Science – which would serve to keep the organization in town.[40] Chancellor Day called the plan to remove the GSC museum "a gross violation of faith," and insisted that no one was as qualified as Dawson "to fight the battle" against it.[41] But an act of Parliament of 1877 swept these misconceptions away, decreeing that the GSC must move to the new national capital.[42] Dawson accordingly determined to mobilize opposition in Montreal until the actual move in 1881.

First Dawson recruited influential friends, including Senator Thomas Ryan and Thomas White, to persuade the government to reverse its decision. The mayor of Montreal, Jean-Louis Beaudry, offered his "hearty cooperation," as well as his own office to hold meetings for planning strategy. A Montreal deputation petitioned Prime Minister John A. Macdonald and the governor-general, the Earl of Dufferin, while Montreal's Board of Trade, City Council, and

Corn Exchange all remonstrated against "the evil." When the lobbying failed – predictably, to those who saw the increasingly powerful hand of Ontario in the whole affair – Dawson tried to salvage what he could. He proposed that a branch museum of the GSC be maintained in Montreal to preserve the original exhibits arranged by the first director, William Logan, and the survey's paleontologist, Elkanah Billings. He suggested that only the more recent collections made during A.R.C. Selwyn's directorship, as well as all duplicate specimens, should go to Ottawa. If this plan failed, Dawson wanted the duplicates for McGill. He even argued that the most precious objects should be left for safety in Montreal; that they might be damaged, ruined, or lost during the move itself or later in the rickety old hotel purchased for the GSC in Ottawa.[43]

Much bad feeling resulted from what was seen in Montreal as the federal government's "want of faith." The promise made by Defense Minister L-F-R. Masson for the creation of a Montreal geological museum remained unfulfilled, as well as a vaguer pledge to leave duplicate materials behind. Like the politicians, members of the GSC expressed little enthusiasm for Dawson's schemes, telling him that he might expect to receive only a small number of specimens, and at some time after the move to Ottawa. Apparently the men lacked both time and money to inventory and identify their duplicates, causing Dawson to complain that "more than enough are rotting in boxes in the Survey Museum ... to remain useless in cellars for years."[44]

In the end, although McGill acquired a magnificent new museum with Redpath's bequest, it lost the national collections that had contributed so much to Montreal's stature as a centre for the pursuit of natural history sciences. At a personal level, Dawson's professional and personal ties to the GSC were so strong that collegial relations were soon restored. But for the residents of Montreal, the loss of the GSC museum came to symbolize the federal government's minimization of their desires, expectations, and assumptions. As Dawson explained to the governor-general, the removal of the GSC museum constituted "a serious blow to science and education in this province, where neither is by any means too well provided [for]..."[45]

EDINBURGH LOST AGAIN: THE PRINCIPALSHIP

In the wake of Confederation, Dawson found himself isolated in Montreal as never before. Earlier he had been conscious of his position as a member of a besieged minority in Quebec; now he felt abandoned by both federal and provincial governments. Once again, he decided to submit his candidacy for an Edinburgh post, this time the principalship. He had accomplished much during his fourteen years in Montreal, but he felt frustrated by the foreign (that is Roman Catholic, whether French or Irish) environment and fatigued by overwork; he longed for a position where there would be "less teaching and red tape and more science."[46]

Once again his bid was defeated. In *Fifty Years,* he claims to have made "no active canvas," perhaps because to Dawson – now a mature educator and scientist of international stature "in the prime of life", as Lyell put it – the rebuff seemed more insulting than before.[47] He was to reflect thereafter on the possible creation of a geology professorship at Edinburgh, a position which subsequently went to Archibald Geikie in 1871, as follows:

I have quite given up any thought of looking to Edinburgh. I could not very well leave my position here for a new professorship without disparagement to McGill and knowing Edinburgh as I do, I would not wish the struggle of getting up a geological class there, without the power and influence of a higher position to work from. I had hoped to have been the means of doing in Edinburgh what Forbes did not live to do, but they have refused me twice; and that cancels all obligation on my part, so they shall not have another offer.[48]

Nonetheless in 1868, despite his denials, Dawson's sense of duty compelled him to launch a respectable attack on the Edinburgh principalship. Moreover, had there been any hope of obtaining the geology professorship over Geikie – the incumbent hand-picked by Roderick Murchison, the chair's founder and Geikie's mentor – Dawson might well have made a third attempt.

Already in early March 1868, following the death of Edinburgh's principal, David Brewster, Lyell was busy assembling support for his protégé and calling in his social debts. He reminded the Duke of Argyll that it was due to his support of George Allman that Dawson had lost his earlier bid for the natural history chair. He hoped that the Duke might now reward the better known and more widely published Dawson with the support that he so richly deserved.[49] He wrote to one of the university curators, David Milne Home, that, as he had now known Dawson for "more than a quarter of a century," he could attest to his "very superior talents, industry, and character."[50] Another old geological ally of as many years, William Logan, praised Dawson's scientific accomplishments and his work in elevating the academic quality of McGill; in both domains, Logan claimed that Dawson had done much to "spread sound science in the colonial dominion."[51]

Despite this strong support, it appeared that Argyll would favour the eminent physician, James Simpson. At least three other distinguished Edinburgh professors were also vying for the chair: another physician, Robert Christison (who, like Simpson, had attended the Queen), a leading surgeon and government consultant James Syme, and the chemist and politician Lyon Playfair. They were joined in the competition by the Glasgow physicist, William Thomson, and Alexander Grant, a classicist. With Baronets (whether hereditary or recently named) Simpson, Christison, and Grant pitted against each other, pundits dubbed the contest the "battle of the baronets."[52] As the Edinburgh writer and literature professor David Masson gossiped to the Scottish physiologist William Sharpey, just "fancy the complexity," in the months to come, and all the "talking, working, and counterworking."[53]

Early in May, Milne Home, as secretary of the Board of Curators, wrote to Dawson with a request for additional information, and told him that they had deferred making an appointment until 18 June, the deadline for filling the chair of moral philosophy. He inquired about Dawson's familiarity with "any other department of science, literature, or art" besides geology, as the Curators should prefer to appoint an individual of "more general acquirements." He asked probing questions about Dawson's social graces – particularly as they might affect his ability to garner support for the university – and about his religious affiliation, adding that the curators expected to appoint someone holding "the Presbyterian and orthodox views of the Community of Edinburgh." He also wanted more detailed information about Dawson's degrees and his duties as principal of McGill.[54] Clearly, knowledge of Dawson's accomplishments and abilities had not spread as widely as he or Lyell might have hoped.

Dawson took Milne Home's questions to heart, soliciting the support of a number of "influential people of Montreal" and elsewhere to vouch for his good character. The Montreal merchant J.B. Greenshields, then resident in London, wrote that the benefits Dawson had afforded "to the cause of education in Montreal" could not be told in a short letter, but that he had brought McGill from a languishing state to an influential position. Dawson, he said, was sociable and hospitable, not a "diner-out," but a "water-drinker." His personality, although conciliatory and judicious, was firm. Despite Dawson's Scottish ancestry and Nova Scotia birth, noted Greenshields, he was "a little Americanized in appearance and accent." Greenshields hastened to add that he did not mean by this that Dawson's manners were "objectionable."[55]

Dawson also contacted the Edinburgh clergyman Donald Fraser (who had resided in Montreal for many years) for assistance, fearing that his scientific backers might have "no good odour" with the Christian people of Edinburgh. On the other hand, he hoped that

Fraser might inform him of the "denominational connections" of the curators.[56] Even more strongly than Greenshields, Fraser emphasized that McGill had been in "a wretched state" when Dawson – "singularly capable and successful as an administrator" – took over; Dawson's "wise, skilful and sturdy management" had raised it to its present position of "influence and prosperity." He spoke of Dawson's affable and considerate manner, and added that Montrealers held Dawson's "evening conversation and music parties" in high esteem. He noted that Dawson's religious tolerance and courteousness were established by the position he held with "universal approbation" in such a heterodox community as Montreal. He possessed neither extreme opinions nor eccentric habits, observed Fraser, and would decline to meddle in church controversies or political strife.[57]

Numbering, of course, among his "scientific backers," Lyell answered Milne Home's queries point by point. He described Dawson's "persuasive manner and gentlemanlike appearance" as a speaker at the British Association for the Advancement of Science meeting in Birmingham three years earlier. He argued that Dawson's publications "sufficiently attest his literary qualifications"; his *Archaia*, for example, showed that he had long studied Hebrew. Although Dawson's theological views led him to conclusions that diverged from his own, added Lyell, he had never known a more conscientious man. He pointed out that Dawson had been principal of McGill for more than ten years, during which time his accomplishments spoke of both his popularity and strong character.[58] Another scientific backer, geological survey director Murchison, echoed Lyell's praise, adding that Dawson's appearance was likely to please "even the most fastidious critics of the Modern Athens." He noted, too, that Dawson's immense knowledge of the Devonian flora alone was likely to cause "a 'furore' in the metropolis of the 'Old Red Sandstone.'"[59] (Murchison thereby associated Dawson with the memory of the popular minister, publisher, and geologist, Hugh Miller.)

Dawson, himself, responded to Milne Home's queries by having private testimonials forwarded confidentially to the curators, and by printing up others for wider distribution: attestations of scientists, politicians, clergymen, and academics. Additionally, Dawson sent a letter (different from the formal application printed with the testimonials) and penned a memorandum. These documents show how earnestly he sought the Edinburgh principalship; they also provide a picture of Dawson "in the prime of life."

In essence, Dawson portrayed his fourteen years of experience at McGill as an "apprenticeship" for the Edinburgh position, "more especially as here the duties of a Principal are necessarily of a much more varied character than in the mother country." Even allowing for "the peculiarities of a colony," he observed that McGill University was "similar in its range of operations to that of Edinburgh." Dawson also described in detail his financial circumstances, in order to show that the Edinburgh principalship would mean a monetary loss for him (around £200), unless coupled with the income from a geology chair, a position yet to be created.

Dawson sent a copy of his *Archaia* to demonstrate his knowledge of Hebrew, although he admitted to little training and cultivation of other ancient classical languages. He even enclosed copies of invitations to his house parties, although he confessed to avoiding "gay society" unless dictated by university interests. He added that his reputation as a public speaker meant many "demands on my time more flattering than convenient."

He digressed to describe the sectarian divisions of the Presbyterian church in Canada, and explained that he followed his father in belonging to the "Canada Presbyterian Church," not the "Presbyterian Church of Canada" connected with the Kirk of Scotland. Nonetheless, he argued that from a religious point of view he would surely be "unobjectionable to the great majority of the clergy and people of Scotland." In his present position, he noted, he had been required "to mingle on intimate terms with persons of all creeds."[60]

Despite his vigorous attack on the principalship, Dawson confided to his daughter that he doubted he would succeed. The decision of 18 June was postponed yet again, until 6 July, probably due to the convoluted wrangling taking place in Edinburgh. There was much "party and fuss" going on; four of the seven electors favoured Simpson, despite the objections of a majority of the professors. Dawson speculated that he could only garner, at best, four votes from among the curators, but trusted that "God will do what is best."[61]

Dawson seems to have had little inkling of the hornet's nest that had been stirred up among powerful interest groups in Edinburgh. First, Christison, who claimed to have been hand-picked by Brewster to succeed him as principal, was portrayed alternatively as apathetic and too political. Although his long-time opposition to the Town Council cost him powerful allies in the community, it brought the students to his side. Then Simpson, who was attacked in a petition circulated by the professors, claimed that Christison had forced their colleagues to sign it. In retaliation, Simpson's supporters circulated their own petition on his behalf. Meanwhile, the *Scotsman* came out against both Grant and Dawson, maintaining that "the office of the principal should not be bestowed on an utter stranger unconnected with the University and totally unacquainted with the British University system."[62] The acrimony of the struggle is indicated in the decision of Christison's sons not to publish his narrative account of the contest even after nearly two decades had elapsed.[63]

Dawson's source of information – his brother-in-law, Peter Bell, who would not "leave a stone unturned" on his behalf – reported that the curators were divided. It would have been simply "astonishing" (as Stanley Frost points out) had Dawson been able to go "from Pictou to McGill to Edinburgh." But at least one curator maintained that, apart from the political complexities of the appointment, Dawson was the best-qualified candidate. According to the notes scribbled on the back of one of the copies of Dawson's printed testimonials, the board members were overwhelmed by Dawson's credentials. Among other assets, they were impressed by his "energy and tact," the breadth of his science and scholarship, his "influence with government," and the fact that he was "amiable and religious." Moreover, "he has friends here and no *enemies* – others have."[64]

This last point would seem to have done in the favourites, Christison and Simpson. Indeed, for more than a century, between the regimes of George Baird (in 1793) and the anatomist William Turner (in 1903), no Edinburgh professor was able to rise through the ranks to become principal [65] In the end, the appointment went to Alexander Grant, who like Dawson had served long years in a colonial situation, in his case in India, where he had been vice-chancellor of Bombay University and principal of Elphinstone College. Like Simpson and Christison, Grant could point to impressive political and governmental appointments and connections, although he lacked the "European reputation" that his friends claimed for him.[66]

If the appointment was destined for an outsider, one wonders what tipped the balance in favour of Grant. He possessed a hereditary title and an "Oxford and Indian reputation, with all the high merits involved in that," yet he had been virtually unknown in Edinburgh at the beginning of the "struggle."[67] Rumour had it that more important than his public credentials was the fact that he was married to the daughter of the late James Ferrier, a distinguished Scottish metaphysician and former Edinburgh professor.[68]

As had happened in 1855, several accidents of fate seemed to move Grant ahead of Dawson. Dawson comforted himself repeatedly that it was simply God's will. Certainly, Grant's role in government, as a member of the colonial legislature, eclipsed Dawson's. Dawson had merely been the ally of provincial politicians during the course of his lobbying for the cause of education. Both Lyell and Donald Fraser insisted that Dawson's chances had been hurt, in addition, by the recent death of Sir Edmund Head, formerly governor-general of Canada and a colonial officer who had been a motive force behind Dawson's career as an educator. (News of Head's death had not even reached Nova Scotia when Dawson first made application for the principalship.)[69] According to Lyell, Head would have "zealously" promoted Dawson's candidacy as "one eminently fitted by his administrative powers and scientific knowledge for such a place."[70]

Grant had been trained as a classicist at Oxford; the success of his edition of Aristotle's *Ethics,* first published in 1857, made him an excellent choice for the chair of moral philosophy at Edinburgh, which needed to be filled at that time as well. (Alas for Dawson, this contest predated the creation of the geology professorship at Edinburgh by three years; natural history was still occupied by Allman). Moreover, the second, vastly expanded edition of Dawson's *Acadian Geology,* which would further enhance Dawson's reputation as a geologist, was published just at the end of the competition over the principalship. Had the 700-page volume been in the hands of the curators from the beginning, it might have done Dawson some good.

When David Milne Home returned Dawson's testimonials, he mentioned to Lyell that he hoped Dawson would stand for the anticipated chair of geology.[71] Lyell, however, expressed reservations to Dawson about the advisability of a third attempt at Edinburgh. As much as he despaired over Dawson's frittering away his time teaching a fresh crop of beginners every year in Montreal, rather than concentrating his attention on geological research, he felt that the fees and salary of such a position at Edinburgh would be unattractive to Dawson. Moreover, he expected that Geikie "to whom it has always been an object of ambition" would have enough influence to obtain the geological chair.[72] For his part, Dawson was disappointed that he was deprived of the opportunity to build up a school of practical science at Edinburgh and the chance to compare the geology of Scotland with that of Canada. But without "the power and influence of a higher position to work from" – that is, the principalship – he said he was uninterested in "the struggle of getting up a geological class" at Edinburgh.[73]

Indeed, when approached about the chair a year later, Dawson was even more definite. He explained that it would damage McGill's reputation were he now to stand for a mere professorship at Edinburgh, especially without "other inducements" to "counterbalance" the "descent in title and position." His present salary (in the range of £1500) vastly exceeded what he might expect to earn in Edinburgh. He believed, moreover, that prospects for the future of his children were far brighter in Canada than in Scotland. He could not resist mentioning that the situation would be far different, had he won the natural history chair at Edinburgh in 1854. By now, he would have built a school of natural science there to rival London's, and his students would be filling "important posts throughout Great Britain and the Colonies."[74]

PRINCETON WON BUT DENIED

During the years that followed, Dawson set about remedying the anonymity that appeared to have cost him the Edinburgh principalship (and earlier, the natural history chair). He became increasingly convinced that the "excessive labour" exacted by his position in Montreal might necessitate deciding between abandoning his "public duties" in Canada and relinquishing his scientific pursuits. The possibility of choosing science over administration led him to undertake a few mineral surveys and to give several public lectures in the United States.[75] From this point on, as he began to deliver numerous talks and to pen innumerable articles for literary and religious periodicals, Dawson toyed with the idea of working as a science popularizer.[76]

His long-time confidant and advisor, the English geologist John Jeremiah Bigsby, warned him, however, not to expect to find a good job in Great Britain. He would labour there "as a colonist under many disadvantages," without the support of "family and political influences." Only a professorship or a "headmastership of a great school" would be sufficiently prestigious, said Bigsby, and these were only "moderately paid." Moreover, all the headmasters had graduated from Oxford or Cambridge. Bigsby described how a friend "with *capital* testimonials" had failed to win a £200 (per year) secretaryship when more than 300 had applied; for that matter, he noted, the great Richard Owen earned only £800 and the botanist William Carruthers, £400 (equivalent to what most professors were paid) at the British Museum. Moreover, despite low salaries, the cost of living grew higher with each passing year in both London and Edinburgh.[77]

Across the ocean, in contrast, said Bigsby, Dawson had "talented children to launch out in the world" whose chances seemed better in North America than in Britain. And salaries were higher, especially for "the New York and Philadelphia professors." Bigsby recommended that Dawson keep on the lookout for "the changes that are always occurring," and in the meantime "keep *absolutely silent* in Montreal as to any desire for relief or change," lest the public forget "all your painful *services* of the many years spent in the College."[78]

Just a decade after Dawson had so eagerly sought the Edinburgh principalship, the campaign bore fruit. He was eagerly sought out by Princeton College and Princeton Theological Seminary. The president of the college, James McCosh, had invited Dawson some years before to deliver a short course in geology and paleontology, which would serve to bring him "into connection with our United States Institutions."[79] Dawson seems to have declined the invitation, but he did accept a later one proffered by the president of the seminary, William Henry Green, to lecture on the relation of science and religion. Like McCosh earlier, Green invited the Dawsons to stay at his own home as houseguests.[80] For Dawson, who also spoke at New York's Union Theological Seminary, it became a triumphant lecture tour.[81] Something of Dawson's charm as a lecturer is revealed in the words of a professor at the Auburn Theological Seminary a few years later, when he claimed that no course of lectures during the past quarter century could compare with Dawson's "in their wholesome influence, and at the same time in their remarkable popularity."[82]

Over the ensuing years, Dawson became a frequent contributor to the influential *Princeton Review,* which claimed a circulation of 10,000.[83] Reprints of Dawson's articles found thousands of purchasers, as well.[84] This exposure in a leading literary periodical gave Dawson stature outside Canada and established him as an informed commentator on contemporary intellectual issues. The future president of Cornell University, Jacob Schurman, approached Dawson to obtain a recommendation from him, a "valued contributor" to the *Review.* He believed that Dawson's word would be "potent enough" to "open the door" to his own contributions.[85] On one occasion, the editor, Jonas Marsh Libbey, asked Dawson to try his hand at providing an original improvement on the respected American naturalist Joseph LeConte's "Man's Place in Nature," which the journal had earlier published.[86] How perfectly Dawson's outlook cohered with the Princeton milieu is revealed in a letter from the director of the seminary, E.J. Craven, who claimed that the views expressed in his own *Princeton Review* article were so similar to those in Dawson's *Origin of the World,* that he wanted to dispel any notion of plagiarism.[87]

In 1877, these initial efforts as a lecturer and writer came together to advance Dawson's interests. At his request, McCosh arranged for George Mercer Dawson to receive an honorary doctorate from Princeton, even though he had been opposed to their "indiscriminate granting."[88] Then in 1878, with the stage presumably set for Dawson to view the proposal in the most favourable light, McCosh made Dawson an offer that he hoped he could not refuse. (Even this occurred in the context of Dawson's writing to him about an engineering professorship for another son, William.) Arnold Henry Guyot, then professor of both geology and geography at Princeton, wished to relinquish the geological portion of his chair. The trustees were willing to hire someone else to take charge of paleontology and geology, but, as McCosh confided to Dawson, "They have set their hearts on you." If Dawson were to give them but the slightest encouragement, he could obtain the appointment.

According to the scenario sketched out by McCosh, Dawson would only have had to teach a few hours each week on specified days, leaving him ample time for original research and travel. Moreover, an anonymous New York merchant offered to underwrite the cost of a scientific westward expedition the following summer, and Princeton offered a salary of $3000 in addition to a house. McCosh apologized that the trustees could not promise a higher salary – he said they could not offend the other professors – but explained that the Princeton Theological Seminary would pay Dawson an additional $1000 annually to deliver a course on the relation of science and religion. (This would have made Dawson's salary equal to McCosh's.) Moreover, added McCosh, Dawson would find life at Princeton less expensive than that in Montreal and with fewer distractions, for there was "good society but no extravagance." McCosh closed with the most persuasive language he could muster, sure to appeal to the missionary spirit in Dawson:

We give you this call. It looks to me as if it were a call from God. It is unsought on your part. There is a concordance of all parties in the place. Your influence will reach over an immense body of young men in the College, in the school of Science, and in the Church. Your sphere of usefulness will have no bounds except those imposed by your strengths.[89]

McCosh offered to meet Dawson anywhere, even in Montreal, to discuss the details. His letter was followed four days later by an invitation from William Henry Green, president of the Princeton Theological Seminary, to accept a permanent lectureship there. Dawson would be free to deal with the range of physical sciences (not just geology, but even anthropology) and their bearings on the Bible. In the view of the faculty, Dawson was an ideal person to teach young candidates for the ministry, for he was able to combine "accurate scientific knowledge with a profound reverence and a strong attachment for the Westminster Confession."[90]

Another colleague at Princeton, Charles Hodge, presented a list of reasons why Dawson should accept the university's offer. Princeton enjoyed affiliation with the Presbyterian Church in the United States, the largest and most influential Presbyterian body in the world. Clearly alluding to Dawson's situation at multidenominational McGill, Hodge added that all of Princeton College's presidents and trustees had been Presbyterians. (Since Dawson could still remember the struggles among the Presbyterian schismatics in Nova Scotia, this point may not have been so persuasive.) Furthermore, said Hodge, the United States needed individuals such as Dawson: "scientific men who are firm believers in the Bible." Anticipating Dawson's response, Hodge noted: "You cannot be more needed in Montreal than you are here." His final argument was that Princeton's closer proximity to the equator (than Montreal's) would surely appeal to Dawson's wife.[91] Guyot also extended a warm welcome to Dawson, assuring him that "science will not lose the advantage of your labors by a change of scene."[92]

Before two weeks had elapsed, McCosh had resumed his assault on Dawson. If Dawson declined Princeton's offer, expostulated McCosh, where else could he go to seek "a geologist of repute who is not a Darwinian." It was not simply a matter of the university's welfare, said McCosh, but of the United States, which needed Dawson "to guide opinion at this critical time." McCosh explained that "vast consequences" followed upon Dawson's decision; he might exert greater influence over public opinion at Princeton than in Montreal and he could contribute stability to the continent. At Princeton, Dawson could define his own field of specialization and help to shape that university's school of science.[93]

Dawson's benefactor, Peter Redpath, was one of the first to surmise that he had turned down Princeton.[94] By mid-April 1878, Dawson sent word to McCosh, Green, and Hodge that until "no hope remains that the Gospel and the light of knowledge can conquer our French Canada," he would be unable to leave Montreal. Dawson explained that his ties to Montreal were numerous: besides his family, he could not desert friends who have "struggled by my side for twenty years," nor his "large collection almost immovable."[95] Even the success of "our Canadian Dominion," added Dawson, depended upon his willingness to stay and aid the course of liberal education, science, and religion.[96] McCosh maintained his hopes over the next several months, emphasizing to Dawson that the unique cooperation between college and seminary had only come about because of his (Dawson's) preeminent qualifications. If he declined, added McCosh, we have not "the dimmest idea of what we may do."[97]

In the end, Dawson turned away from Princeton's pleas. Instead, he looked ahead to the bright future in Montreal symbolized by Redpath's promise of a new museum to house his natural history collections. One may well wonder what caused him to reject the call, especially as it seemed to answer his plans and prayers of the previous decade. One point should not be forgotten, however: Princeton offered not a presidency, but merely a professorship (albeit well-endowed). Furthermore, despite the annoyances in Montreal, it was a large metropolis, whereas Princeton must have seemed a provincial backwater. Dawson clearly did not want to leave Montreal or Canada, the country of his birth. Nor did he want to give up his varied activities as an educator and science booster in order to escape into a world of pure scientific (or religious) abstractions, however often he made assertions to the contrary. Perhaps he remembered a plaintive, homesick letter that he had received from Thomas Sterry Hunt a few years before. Hunt, then in Boston at the Massachusetts Institute of Technology, yearned to return to Montreal, where he had lived for twenty-five years; he was simply "too old a tree to transplant."[98] Dawson, too, had sunk his roots deeply into the alluvial deposits on the edge of the Laurentian shield, despite the presence of rocky outcrops that impeded the development of many of his dreams.

7 "None knew him but to love him"[1]

When Dawson declined Princeton's call, he cited the brighter prospects that Canada held for his children's future. This implies that he saw his own life in terms of sacrifice for his children; indeed, this attitude is explicitly conveyed elsewhere in Dawson's papers. Such a conclusion is only partially true, however, for it tends to distort the relationship between father and offspring. Although Dawson strove to advance his children's interests as he understood them, he never hesitated to bring their talents to bear on his own activities. His strong convictions, unwavering sense of purpose, and zealous prosecution of his responsibilities exacted a psychic toll upon the Dawson household. This darker view of Dawson's parental role is succinctly expressed by his youngest son Rankine: "Everything had to take a second place (even the welfare of his own family) where it interfered in any way with the perfecting of his own life and work."[2] Indeed, the close friendships sustained by Dawson might seemingly have been nurtured for their ability to fuel his ambitions and further his aspirations. This said, it is unlikely that Dawson was aware of the personality trait. Nor does this characteristic lessen the real affection he felt for his family and friends.

THE DAWSON CHILDREN

Certainly, like other middle-class fathers of the day in Montreal, Dawson took an active role in the lives and rearing of his five children.[3] Eldest daughter Anna recalled years later the wonderful Saturday morning excursions made by the children and their father throughout Montreal and the surrounding countryside. Sometimes they visited commercial establishments, whether the post office or a printing company; on other occasions, they participated in geological and botanical excursions. They were always expected to dine with their parents, as well, which constituted a "great education for us." At the table, their father related jokes, odd stories, and quotations from (as the children were to learn later) the writings of Lord Byron, the *Pickwick Papers* of Charles Dickens, or the novels of Sir Walter Scott.

Anna describes a serious and industrious father, who nevertheless indulged the children's monetary requests (their mother, on the other hand, exercised Scots parsimony). She fondly remembers his gift of a beautifully illustrated *Arabian Nights,* which "illuminated that period of our childhood." He possessed a quick temper, although he rarely let it show; and he had no interest in games or music, which he saw as hindrances to more serious enterprises. In his opinion, only drawing functioned as an acceptable avocation; thus, he often tried his hand at sketching places he had visited, or geological and natural history specimens he had examined. One of his few recreations was gardening. He was seen on more than one occasion "with muddy boots and turned up trousers measuring out and tracing the form for garden beds" on the McGill campus. He planted exotic and diverse varieties of trees (many sent from Pictou), shrubs, vegetables, and flowers, and planned old-fashioned perennial beds that looked especially attractive by the "pretty little brook that flowed through the College."[4]

In short, Dawson was a loving, devoted father, yet someone who unswervingly pursued "the light given to him."[5] Son George recollects that the children seldom saw any "flash of the inner man" who tended to erect mental fortifications against others. But he was also as private and introverted as he was serious. He used the written word guardedly; as George puts it, "hazarding nothing in open speech."[6]

In Dawson's case, parental solicitude was tempered with an unwavering conviction that his male children, in particular, should find practical careers that usefully complemented his own interests. He never hesitated to solicit his children's assistance, even asking them to translate letters from European correspondents.[7] Elder daughter Anna, who for years acted as his literary secretary and illustrator, found escape in marriage to chemist Bernard James Harrington (although Harrington's mineralogical skills made the alliance exceedingly attractive to Dawson). In fact, the only child who seems to have escaped impressment into "Papa's" service was Eva, perhaps because she was a girl and so much younger than the other children. Without a clear niche in Dawson's program of familial cooperation, she has virtually faded from historical notice.

Always quick to compensate for any of Dawson's shortcomings was his wife, Margaret Mercer, who devoted herself to looking after, and fretting over, the welfare of their children (although, according to her own granddaughter, her husband always came first).[8] Their first child, James Cosmo, died as an infant in Nova Scotia in July 1849, less than a month before the birth of George Mercer Dawson. Another sad blow for the young household (already expanded by the birth of Anna Lois in 1851 and William Bell in 1854) came when George contracted a life-threatening disease at the age of nine. Although he survived the ravages of tuberculosis of the spine, or Pott's disease, it left him with "the stature of a ten-year-old" and "the bulky torso of a hunchback." His mother was so distressed by his appearance that she tended to protect him by hiding him away from the scrutiny of others, even her closest friends.[9]

Whether due to the accident of his position in the Dawson family (as the oldest child and the eldest son) or to his determination to overcome the limitations of his handicap, George Mercer Dawson spared no effort to become the perfect, devoted offspring. In 1869, at the age of twenty, he embarked on a geological career by enrolling in the demanding program at the Royal School of Mines in London. He complained to his father about the "squashiness" of dissecting a scallop and the uselessness of remembering the formulas of organic chemistry.[10] His academic performance improved remarkably over the course of his three years there and, ultimately, he graduated first in his class with a prized collection of awards, scholarships, and honours collected along the way.[11]

From the beginning of his stay in London, George assiduously attended to the needs of his father. He regularly sent gossip about his father's professional colleagues and friends, obtained fossil specimens, and forwarded news of the proceedings of the Geological Society of London and the contents (as well as copies) of scholarly journals. He provided evaluations of laboratory instruments, and helped to advance Dawson's intellectual interests and publishing concerns. It was George who sent his father a copy of the first issue of *Nature*, explaining that many celebrated scientists numbered among its contributors.[12]

Having matriculated at one of the world's most prestigious schools of science and technology, George felt qualified to pass judgment on his father's attempts to establish a school of applied science at McGill. He freely gave his opinion concerning candidates for professorships, curricular matters, and the direction the development of the school should take. As he himself approached graduation, he contemplated a mining career or a role with a geological survey; in either case, he hoped to return to Canada.[13] The British survey had already mapped most of the interesting territory within its purview; moreover, pay was small and "advance almost hopeless," given the limited number of senior positions.[14]

George's voyage home from London was paid by a British investor, in return for an analysis of iron ores near Pictou. George learned shortly thereafter that this job conflicted with his father's consulting work for another company.[15] Nevertheless, George's work proved to be a fruitful source of information for the elder Dawson: it allowed George to report to his father about the activities of his business partners and to spy on his father's competitors. George also alerted *père* Dawson whenever interesting properties for investment and eventual exploitation presented themselves.

George subsequently found employment as a naturalist and geologist to the North American Boundary Commission. In 1875, he was appointed geologist to the Geological Survey of Canada, where he worked his way up to become director in 1895. This was the position he had sought above all others, and it was to offer him everything he had ever dreamed of.[16] His father, who at first had persuaded him to decline a position on the survey in favour of the boundary commission, made an uncharacteristic about-face and energetically worked to secure his son's appointment.[17]

George's physical limitations were so severe that his father's close friend, John Jeremiah Bigsby (who acted as surrogate father to George in London), worried that fieldwork might kill him.[18] But despite his disabilities and the formidable obstacles against exploring western Canada at that time, George found in the Canadian west the opportunity to become perhaps the most brilliant field geologist the country has ever known. His arduous geological campaigns seemed almost calculated to defy the limitations of his crippled constitution. His finely honed intellectual tools permitted him to forge his observations into powerful and challenging synthetic statements. He contributed actively, for example, to the debate over continental glaciation, defending the perspective of his father for many years.[19] Almost a footnote to his larger geological program, the extraordinary ethnological data collected during the course of his survey work thrust him into prominence as one of the pioneers of Canadian anthropology.[20] Here again, George's interests complemented those of his father, who had earlier studied the language of the Nova Scotia Micmac Indians and continued his anthropological investigations in Montreal.[21]

Undoubtedly, it remained a source of regret for Dawson *père* that the son with whom he shared such a close intellectual affinity (and with whom he often collaborated, if only by post) should spend so much of his adult life separated from home by an immense continent. To some extent, the father saw the survey as squandering the remarkable energies of his delicate son; on one occasion, he remarked that it was a pity that George did not turn his "power of writing into more popular and widely useful channels than Reports."[22] In an effort to bring him back into the sphere of the east coast North American establishment, the elder Dawson arranged for Princeton to bestow on George an honorary doctorate.[23] He was quick to alert him whenever it appeared that a suitable teaching position might become available, whether in Ottawa, Kingston, or Toronto.[24]

For his part, wherever George's travels took him, he always looked out for his father's special interests. Unusual fossils and other natural history specimens that he sent back from western Canada found their way into his father's hands, even though most of them eventually wound up in the survey museum (itself located in Montreal until the move to Ottawa in 1881). Nor did he hesitate to acquaint his father with coal-mining prospects in British Columbia.[25] Several trips to Europe during the early 1880s offered additional avenues for advancing his father's scientific interests.

Closest to George in age and psyche was younger sister Anna (later she reflected that they were both, by far, the most like their father of all the siblings).[26] She seems to have been raised with the purpose of assisting her father, who especially encouraged and instructed her in "sketching from nature." Indeed, she supplied wood engravings and other illustrations for his books and articles.[27] Even George maintained that Anna could work with "Papa" and help him "in a way none of the rest of us could."[28]

Anna's range of duties included helping her mother answer her father's mail. Her own letters, in turn, became a precious commodity for Dawson during his tours away from Montreal, especially because she never hesitated to give her views about their common acquaintances. She condemned Thomas Sterry Hunt, for example, for his disgraceful treatment of his estranged wife, but called the Penhallows "a decided success" in their new position among the McGill faculty.[29] To Dawson, her forthright opinions were invaluable, for she gave him "much more of college and other gossip than [brother] William condescends to give."[30]

On 7 June 1876, Anna married the modest and self-effacing Bernard Harrington, a former student of her father's who had returned to Montreal after earning a doctorate at Yale University's Sheffield Scientific School. Harrington became a lecturer at McGill in

1871. In characteristic fashion, he claimed not to be unhappy with his small salary since "I am not worth any more," and said he did not expect to become a "man of means."[31] (His lack of ambition and self-esteem appears to have been partly responsible for his future family's existence in a state of genteel poverty.)

The wedding took place in the Dawson drawing room, decorated with flowers undoubtedly collected from the McGill gardens: "wreaths of lilac, Siberian honeysuckle and cherry." The table displayed no "hideous erections of barley sugar," but only "quantities of lovely flowers," all white, including "a whole row of specimen glasses down both sides ... filled with lilies of the valley." Harrington's groomsmen were his colleagues from the geological survey, director A.R.C. ("Cecil") Selwyn and Robert Bell. The wife of McGill's secretary and registrar, William Baynes (whose family's residence adjoined the Dawson's), played the piano. She also gave Anna the lace with which to trim her cream-coloured silk gown imported from Scotland. Mrs Redpath's "grandest carriage" took the newlyweds to the train station after the wedding, which had been attended by associates and close friends from McGill and Montreal: the Campbells, Cornishes, and Murrays, as well as the Redpaths, Molsons, and Macdonalds.[32]

Even before the glorious wedding day, Dawson seems to have taken Harrington's interests to heart; perhaps he realized that he would soon be a member of the clan.[33] George's earlier jealousy of Harrington aside, his father helped Harrington to secure a position on the survey and then smoothed the path towards a professorship (and eventually an endowed chair) at McGill. Although he was eager to propose his son-in-law's name for membership in the London-based geological and chemical societies, Harrington protested that "too many letters after one's name are troublesome and would certainly swamp me."[34]

Certainly the relationship between the Harringtons and the Dawsons became an especially close one. The elder Dawson built a house for the newlyweds on property adjacent to the McGill campus (which he had earlier tried to persuade the board to purchase for the university). They summered together at Little Métis (the Dawsons in Birkenshaw, the Harringtons in a more modest cottage next door, purchased for them by Anna's father). Family photograph albums show that the nine grandchildren produced by Anna and Bernard provided a never-ending source of delight for William and Margaret. When William died in 1899, the Harringtons cut a passageway between their house and the adjoining Dawson home (by this time they had left the McGill campus for a University Street residence), in order to facilitate communication.[35]

Like his older brother George, second son William Bell Dawson also ventured to Europe for advanced scientific training. At the age of twenty-one, with McGill arts and applied science degrees in hand, he earned acceptance to the prestigious École des Ponts et Chaussées in Paris.[36] During his three years in Paris, he distinguished himself as a model student and even won a scholarship from the Institute of Civil Engineers. Like George, he was always willing to carry out tasks for his father while abroad, however mundane; for example, he helped to examine, and arrange the purchase of scientific apparatus and instruments for McGill.[37] He used the opportunity of being in Paris to find meaningful summer employment (which both fulfilled his obligations to the École and would later serve him well in Canada), as well as odd jobs, such as technical writing.

William assured his father that he placed the securing of "settled employment" above every other consideration. Indeed, he said he would prefer to give up his profession entirely rather than relinquish this quest. Although he now viewed the position of professor, in particular, as nearly "the acme of perfection as an occupation," even Dawson's close friend Daniel Wilson was unable to secure a chair for him at the University of Toronto.[38] He, nevertheless, assiduously applied for a range of engineering positions, including the position of assistant engineer to the Quebec Harbour works and head of the engineering division for the Grand Trunk Railway. He always insisted that virtue was its own reward; he cared little for the prestige bestowed by the initials after one's name and insisted that "a good name is better than riches."[39]

William seems to have been defeated by his humble and almost ascetic attitude. He wanted to publish his description of a machine for calculating latitude, for example, not because he wanted a patent or to stake a claim "in the teeth of clenched antagonisms," but because it might enhance his reputation.[40] Even his job search seems to have been doomed by his quest for a position that would allow him to realize his potential as a civil engineer to the utmost. His father tried to persuade him to apply for a professorship of mathematics, natural philosophy, astronomy, and engineering at King's College (Nova Scotia), but William interpreted the job as one dedicated to teaching pure mathematics, and therefore detrimental to his professional advancement.[41] (Later, however, Anna was to maintain that he had never been given a free hand and the support to do his best.[42])

Eventually, William was hired to work on the topographical survey of Nova Scotia; at least this allowed him not to relocate to the Rocky Mountains – a prospect that he viewed with abhorrence, unlike his two brothers – although the job appeared to be more closely connected with geology than civil engineering.[43] When the Nova Scotia government decided to disband the survey, William reported with pride that he would not "condescend to any lobbying in the matter."[44] He then went on to work as a civil engineer for the Dominion Bridge Company. But even there, William complained that if he wanted to advance and attract the attention of his superiors, he would have to do "some dishonest and unreasonable thing," which was why he preferred to hold back. Reflecting on his father's work at McGill, he somewhat uncharitably suggested that building up an institution was "a more or less uncertain undertaking," and that societies and organizations had little use except in so far as they brought their efforts to bear on an individual.[45]

As William approached age thirty, his father seemed to despair of this behaviour so unlike his own, especially as manifested in William's incapacity to find permanent employment. Apparently when pressed about this issue, William philosophically responded that "the question of what one *is*, is of so much more importance than of what one *has* or *does*," that he could hardly contemplate his future actions.[46] His mother was so vexed by his lack of job prospects that she berated her husband for not taking "sufficient advantage" of their acquaintance with such people of influence as William Cornelius Van Horne and George Stephen on William's behalf.[47] Finally, in 1884, he was appointed assistant engineer to the Canadian Pacific Railway. A decade later, he became director of the Dominion Survey of Tides and Currents.[48] Ironically, as a mature scientist, his interests turned to writing on evolution; like his father, he opposed the theory.[49]

In contrast with the steadfastness of purpose exhibited by William and George, younger brother Rankine found himself constantly distracted from the task at hand. As his niece would describe him in later years, he seemed to be stuck on a landing between two flights of stairs, neither going up nor down. It is interesting to speculate about what happened to Rankine, who seemed to his sister to be perhaps "the most promising of us all."[50] At her wedding, for example, he had been indispensable, and a friend christened him Mercury for having run so many messages.[51]

Rankine displayed an entirely different personality from those of the rest of his siblings. Older sister Anna saw him as "set in an absolutely different key." She wrote of his "distorted vision" (rather than the bad luck that Rankine claimed to have encountered) and wondered whether a "brain fault" were at the root of his checkered career, peripatetic ways, and difficulty in getting along with others.[52] Eva, who as the youngest was his companion as a child, confided that she thought his judgment had been imbalanced by being lost in the woods during a trip to the west; ever since, she said, he seemed to be a completely different character with "peculiarities of manners and ways."[53]

Part of Rankine's problems may well have stemmed from his position in the Dawson household; even in his early twenties, he had to remind his father that although he was his third son, "I cannot on that account always remain a boy." Sent to New York City to earn a medical degree in the early 1880s, Rankine was forced to report his failure, but nonetheless insisted that he had learned something. He stood in awe of Fifth Avenue, with "its innumerable luxurious carriages, and endless stream of handsomely dressed people." He reflected to his father that if only he had seen "the prizes which commercial life has to offer" at an earlier age, no other pursuit would have held any attractiveness.[54]

Although apparently his experience in New York brought Rankine to a state of mental collapse, he did become a member of the Royal College of Physicians in London shortly thereafter.[55] In the intervening few months, he found time to dabble in real estate in Manitoba (where he believed his father had acted counter to his own interests). Although now he was only "a penniless student, without standing or prospects," he said he had been "an amateur financier" all his life, and prided himself that "if I understand anything it is Manitoba and our North West Territory." Rankine's mercenary, materialistic traits contrast sharply with the selflessness and generosity of his parents. He claimed to have accepted money from his father on one occasion not because he needed it, but because he believed that it was safe in his hands "from the ravages of museums, societies, and other hungry monsters who are in the habit of preying upon your purse."[56]

Signing on as a surgeon with the western division of the Canadian Pacific Railroad seemed to offer Rankine scope both to "make money" (which he assured his father was "a good thing," contrary to what he had been brought up to believe) and to satisfy his parents' expectations about a proper career. But even he was unprepared for the roughness of frontier life and the absence of the amenities of civilization that he had taken for granted. He left the railroad to serve as a surgeon on the liners of the P & O Company, which took him to Asia, Australia, and Europe.[57] After four years in this capacity, he married and settled in London, an ocean apart from his parents and siblings.[58]

DAWSON'S CLOSEST FRIENDS

Dawson's success, though it was not achieved without cost to his immediate family, can be explained largely in terms of his family's unquestioning devotion to him. That loyalty, however, could be rendered very differently, whether by the headstrong, unpredictable Rankine or the fiercely independent George. As well, apart from his family, Dawson came to depend on a few close professional colleagues. The importance of these ties of collegiality were revealed on numerous and frequent occasions. Charles Lyell, for example, acted on behalf of his protégé continuously, whether defending Dawson's interests amid London's scientific societies or abetting his strategy during competitions for the Edinburgh chair and principalship. The English paleobotanist William Crawford Williamson commiserated with Dawson over the Royal Society of London's shabby treatment of his memoir on Devonian plants (evidently, the reluctant Royal Society had taken twenty long years to publish his own work on coal fossil flora).

Three other special relationships anchored particular phases of Dawson's career. The first of these was the nurturing support of the English paleontologist, John Jeremiah Bigsby. Although Bigsby lacked Lyell's intellectual preeminence, he did everything he could to promote Dawson's fortunes in scientific circles in Britain. Dawson later thanked God for giving him "the friendship of this excellent and able man."[59] A similar rapport developed over the years between Dawson and the American paleontologist, James Hall, working as they did on parallel problems (separated only by the artificiality of the border) and participating in the same organizations. For both the Englishman Bigsby and the American Hall, a strong Canadian component figured in their paleontological work and activities. (Moreover, Bigsby held Hall in high regard, praising him in his *Thesaurus Devonico-Carboniferous*, for "the breadth, sobriety, and accuracy of his many generalizations, and for the enormous amount of good paleontology he has produced."[60]) Exemplifying a third phase in Dawson's career, during which he moved away from his earlier attachments to foreign sources of prestige and recognition, was his association with fellow Canadian university president Daniel Wilson of Toronto. One indication of the importance of these relationships to Dawson is his dedication of chapters in *Some Salient Points in the Science of the Earth* to both Bigsby ("Markings, Footprints and Fucoids") and Wilson ("Early Man").[61]

Purely intellectual interests seem to have brought Dawson into contact with the Nottinghamshire physician and geologist J.J. Bigsby. A generation older than Dawson, Bigsby had travelled to British North America as early as 1818, first as an army surgeon, and then as secretary and medical officer to the British party of the International Boundary Commission. The commission, charged with drawing the boundary between the United States and Canada according to the terms of the Treaty of Ghent of 1783, gave Bigsby the opportunity to travel nearly 2000 miles across the North American continent. Along the way, he could indulge his passion for geology, which dated back to his days as a medical student at Edinburgh. Returning to England for good in 1826, Bigsby published the fruits of his geological explorations in the

American Journal of Science and the *Transactions* of the Geological Society of London.

Twenty years later, Bigsby decided to move from Newark-upon-Trent to London. The change allowed him to abandon his medical practice and devote more time to his literary and scientific interests. In 1850, he produced a two-volume memoir of his Canadian travels, charmingly entitled, *The Shoe and Canoe*.[62] He also continued to write about North American geological topics until 1864, despite his decades-long absence from those shores.

Possibly because Logan had deemed one of Bigsby's early papers as "the first essay of any importance upon the fossils of Canada," Dawson began to correspond with Bigsby in 1861.[63] Later the intellectual borrowing was reversed, as Bigsby expressed his indebtedness to Dawson in his *Thesaurus Siluricus,* a dictionary of nearly 9000 different species of fossils, which first began to appear in installments in 1868.[64] The Dawson-Bigsby correspondence spans two decades, concluding only with Bigsby's death in 1881.

Although Lyell always functioned as Dawson's mentor, adviser, and defendant in scientific matters – Bigsby called him a "solid and painstaking friend"[65] – it was Bigsby who acted as Dawson's eyes and ears in London, never shirking from transmitting innuendo or rumour, perhaps because he was well aware of Dawson's intellectual isolation in Canada. One of Bigsby's remarks in the introduction to *The Shoe and Canoe* serves to reflect his correspondence, as well: "The impersonal is unreadable; it is the current incident of the day which gives transparency and life. Some may say, that I gossip a little."[66] In his autobiography, Dawson refers to Bigsby as one of his most valued correspondents over the years, not only for sending him "notes of scientific gossip," but for giving him "advice regarding the propriety or expediency of taking part in discussions, or replying to criticisms."[67]

Bigsby was especially zealous about keeping Dawson informed of the affairs of scientific societies. On the eve of Dawson's election to the Royal Society of London, for example, Bigsby wrote that Lyell had had Dawson's nomination supported by "six of the greatest geologists in Britain," including Murchison and Darwin. Bigsby fully expected Dawson to achieve the rare distinction of election on first proposal (as he did, in the event). According to Bigsby, Dawson had earned this honour through his position, works, and personal character.[68]

Perhaps the fact that Bigsby himself did not become a member of the Royal Society of London until 1869, explains why he declined to meddle in the controversy over the society's unprecedented decision not to publish Dawson's Bakerian lecture on Devonian plants.[69] Not only did he feel he could not ask the names of the referees, he believed that it was best for Dawson not to know who they were.⁷⁰ Later, however, he urged Dawson to apply for a Royal Society grant (which, in the end, he did not receive), and to let "bye gones be bye gones" rather than allow "a grudge (seven years' old) [to] forbid doing a good work." Perhaps because he had by then made Dawson's acquaintance (and ministered to George Mercer Dawson's needs during his three-year stint at the Royal School of Mines), he could accuse Dawson of being "perniciously over-irritable."⁷¹

A fellow of the Geological Society since 1823, Bigsby was especially involved in its affairs. He endowed one of the society's medals (its terms dictated that it be awarded to a scientist, under the age of forty-five, examining American geology) at a cost of only £200. By contrast, Murchison had left £1000 and Lyell, £2000, for their medals; but, as Bigsby explained to Dawson, they were "giants both in science and in finance."⁷²

Despite his keen interest in the affairs of scientific London, Bigsby viewed societies such as the Royal and the Geological as fundamentally grasping. They all wanted to be wealthy, no matter the cost to the branches of knowledge they were supposed to foster. They amassed thousands of pounds in their coffers, while they refused to pay their loyal and hardworking employees a decent wage.⁷³ As a result of this "false economy," the officers of most scientific societies were forced to live by their pens. They were reduced to editorial work, producing manuals and helping weak authors with their prose.⁷⁴

Bigsby felt a special obligation to send Dawson "news about geologists"; perhaps, having visited Canada decades earlier, he sympathized with Dawson's lack of contacts overseas. He described "a new set of writers and debaters full of matter" – including John Whitaker Hulke, John Wesley Judd, Andrew Crombie Ramsay, Peter Martin Duncan, Robert Etheridge, Jr, Henry Hicks, and William Carruthers – rising up to occupy the places of the "falling heroes": the "Greenoughs, Murchisons, Horners, Salters, Hopkins, etc."⁷⁵ As the older generation of geologists became "extinct," these ardent and hard-working "new minds" would bring "new light to new subjects" and find "capital and work" in the enlarged staff of the Geological Survey of Great Britain.⁷⁶ Bigsby seemed to delight particularly in forwarding gossip about the surveyors.⁷⁷

Bigsby's second fossil dictionary, the *Thesaurus Devonico-Carboniferus* (1878), dealt precisely with the fossils Dawson had spent so much time identifying and describing. Here Dawson aided Bigsby, sending him, on one occasion, a "magnificent present" of one hundred new Devonian plants, thereby apparently doubling the number of specimens listed therein.⁷⁸ In the end, Bigsby decided to include Devonian fossils along with the Carboniferous, which meant that the work described some 15,000 species. He named Dawson and the late American paleontologist Fielding Meek as the "chief benefactors" of the work, wondering "when shall we see their equals as narrators, as interpreters, and as undaunted explorers?"[79]

Bigsby's keen interest in fossils was complemented by his intense religious fervour. As he wrote in the introduction to the *Thesaurus Siluricus,* "As long as an individual Mollusk remains unregistered it loses a great part of its usefulness in natural history; and we remain ignorant of its place in Creation."[80] In a similar vein, in the *Thesaurus Devonico-Carboniferous,* Bigsby explained that the Carboniferous flora provided but "a single portion, one trace, as it were, of an exhaustless treasury of proof upon proof of Omnipotent design."[81] Such a deistic evaluation of the utility of paleontology accorded comfortably with Dawson's views, as did Bigsby's strong denunciation of Darwinian evolution.

United in their "strong, distinct and trenchant" opposition to the "sad hypothesis" of evolution, Bigsby recommended Dawson as an essayist to the editor of *Leisure Hour,* a publication of the Religious Tract Society.[82] But their common front was not enough to withstand the advances of the powerful new doctrine. Moreover, Bigsby saw an assault on faith everywhere around him. Even at his favourite haunt, the Geological Society of London, he was shocked by the indifference that greeted various "infidel opinions."[83] At the Royal Society, he expostulated, there is "belief in no God, no Bible is openly paraded."[84] In his view, most men of science in the London of the day were but "triumphant boastful materialists and unbelievers."[85] Bigsby comforted himself with the decline in public opinion that the Darwinists' doctrine of disbelief had undergone as a result of the "outrageous denials" of some of their number.[86]

Although the relationship between Bigsby and Dawson (like that between Lyell and Dawson) tended to be paternalistic, perhaps because the English scientist belonged to an older generation, Dawson's letters to James Hall of Albany, New York's state geologist, show an exchange between equals. Their correspondence spans four decades, from 1856 to 1897. Not only did they share a special concern with invertebrate fossils and with the Devonian and Silurian formations, they also brought the same spirit of tenacity and stubbornness to bear on all their investigations. Neither were strangers to controversy. But whereas Hall made his mark as a fieldworker and *doyen* of geological survey work in America, Dawson's talents took him into the halls of academia.

Dawson seems to have met Hall shortly after his arrival in Montreal in 1855; Hall was then identifying fossils for the Geological Survey of

Canada, and Dawson asked him to examine his collections from Nova Scotia.[87] Their common involvement in the activities and meetings of the American Association for the Advancement of Science further developed the close relationship between the two men. It all began in 1856, at a meeting of the association in Albany at which Dawson commandeered the organizers in order to bring the society to Montreal the next year; without Hall's support, the bid would have failed.[88]

Shortly after the Montreal meeting, the two scientists began to exchange fossils, reprints of scientific papers, and opinions. Dawson sent Hall "a carefully selected suite" of lower Carboniferous fossils from Nova Scotia for comparison with those of the western United States (Hall also worked for the surveys of other states, including Iowa and Wisconsin). Dawson intended to make use of Hall's comments in a supplementary chapter to his *Acadian Geology*.[89] In return, Hall sent Dawson his collection of New York Devonian plants.[90] For the next four decades, fossils passed back and forth between the two men, especially crinoids, graptolites, mollusca, and fragments of fossil plants.

In contrast to Hall's dealings with many other scientists, particularly some of the members of the Geological Survey of Canada, Dawson and Hall always shared a cordial, cooperative relationship. It is hardly surprising that competitiveness arose between the Canadian survey (headquartered in Montreal) and the New York survey (directed from Albany), with both organizations examining contiguous rock formations. But a particularly long and vituperative priority dispute with Elkanah Billings, dating back to the early 1860s, marred Hall's relationship with the Canadian survey.[91]

Although Hall and Billings (supported by Logan) sparred over the nomenclature and stratigraphy of some Paleozoic rocks known as the "Quebec group," the argument also related to money. Hall had contributed his paleontological skills to the crowning achievement of the Canadian survey, the *Decades* (Logan's designation for these monographs, each of which contained ten illustrations), and had never been entirely paid for his efforts. Furthermore, Billings, who had been educated in law rather than science, won the appointment as paleontologist to the survey, a position that had been earmarked for Hall earlier. Hall accused Billings of resorting to the cunning tactics of "a low lawyer."[92] He remarked, moreover, that at this time of American strife, it was a great pity that there could not be "men of some profession who can avoid bitterness and reproaches."[93]

Hall left petty politics and bickering aside, however, when it came to his life's work, the publication of his *Paleontology of New York* in thirteen volumes (1847-1894), which brought together the fruits of his geological explorations and reflected the enormous diversity and extent of his collections. As one biographer states, "he knew no duplicates; no two specimens of a species seemed precisely alike."[94] To Hall, anything that interfered with exploring, collecting, or publishing was "a nuisance to a scientific life." He had no love for "the work of annual reports, catalogues, recommendations, etc."[95] Dawson, in contrast, saw these administrative distractions as part and parcel of his calling as a scientist and educator. Increasingly these kinds of mundane duties brought him into closer association with individuals who better appreciated his consummate talents in dealing with a host of demands on his time and energy.

One of these was the archaeologist Daniel Wilson. Wilson arrived in Toronto from Edinburgh as professor of history and English literature at University College in 1853, about the same time that Dawson arrived in Montreal from Pictou. The two men did not begin to exchange letters until nearly 1870, when their common interest in anthropological artifacts brought them together. Their correspondence became particularly intense during the early 1880s, at the time of the creation of the Royal Society of Canada, when Wilson became president of University College.

Wilson never hesitated to ridicule and satirize his contemporaries, whether colleagues at the university or politicians at all levels of government.[96] On one occasion, complaining about the silliness of academic affairs, he reminded Dawson that the University of Toronto had passed over Thomas Henry Huxley as a professor in favour of the totally unknown and unaccomplished William Hincks.[97] Their correspondence appears to end around the time that Wilson became president of the University of Toronto, in 1887, perhaps due to Wilson's increased responsibilities. As he wrote to a sympathetic Dawson in the midst of these new claims on his time, it had been "endless toil and trouble from morn till night."[98]

Usually Wilson and Dawson wrote about educational matters, and their varying situations, experiences, and difficulties in upper and lower Canada. Like Dawson, Wilson was persuaded that universities in Canada must follow a practical course adapted to their milieu, untrammeled by the example of European institutions such as Oxford and Cambridge.[99] Many letters, for example, address the issue of coeducation and the problems raised by educating women at the university level. In addition, the two administrators were united in their belief that political considerations and religious dogma had no place in university affairs. Wilson distrusted politicians heartily, whether they exercised their machinations over local, provincial, or national affairs. He described the Toronto municipal aldermen as "crooked" and given to "treachery and double-dealing"; he called them "an unprincipled set of jobbers, who as they get no pay, and plenty of abuse, feel at liberty to make what they can out of it for themselves and their friends."[100] On another occasion, hearing that Dawson was to be knighted, he wrote that the effect of classifying Dawson and Logan with politicians like Tupper and Tilley, was to make an honorary doctorate or a Fellow of the Royal Society [a DCL or an FRS] "a hundredfold more covetable than a KCB [a Knight Commander of the Bath]."[101]

When appointed president of University College, Wilson, who for years had worried about the covert operations of the Methodists and Presbyterians at the Toronto colleges as they tried to establish control over the direction of the university at large, saw himself as one of the few individuals who might give "a fair trial" to a non-denominational organization.[102] In his view, a man educated in one of the Scottish universities (such as himself or Dawson) "would have a vast deal less to unlearn."[103] Nonetheless, he realized that his problems paled besides Dawson's struggle against "the priests" and the ultramontanism that "will quench the light wherever she is not effectually resisted."[104]

Wilson, like Dawson, distrusted the high priests of science as well as the high priests of religion, believing that "science makes no skeptics." He was, nevertheless, unwilling to do battle with those "skeptical men of science [who] will naturally turn materialistic weapons to account in their own defence."[105] Neither Wilson (the Presbyterian anthropologist) nor Hall (the Roman Catholic paleontologist) found himself ineluctably drawn into the debate over evolution, as were Dawson and Bigsby before them.[106]

In Dawson's case, his sense of religious and scientific "calling" produced a missionary zeal to conquer evolution, which he qualified as crass materialism and atheism masquerading under another guise. Unfortunately, however, this commitment was to ultimately diminish his extraordinary accomplishments as an administrator. In the latter realm, his brilliant initiatives and execution of a range of undertakings placed him firmly among the most progressive figures of the twentieth century. But his strong ideological objections to Darwinism, as well as his firm stance on the wrong side of several other scientific controversies, increasingly made him seem irrelevant to young practitioners intent on exploring new frontiers in the natural history sciences. Although his active role as a scientific controversialist (however much this commitment was minimized in his own autobiographical accounts) fueled the flame of his zeal, it also distracted him from more specialized scientific endeavours, where he might have built a more enduring reputation.

8 "One of the deepest mortifications of my scientific life"[1]

Dawson remained ever conscious of the huge toll that his educational and administrative work exacted upon his scientific creativity. His commitments to McGill and his involvement with other institutions meant that all his writing and research had to be done when he could steal away a few hours from his myriad responsibilities. Even summer holidays at Métis proved to be a mixed blessing: they brought precious leisure time, but also required him to work far from his collection of specimens and monographs. Acutely aware of how his work depended upon "desultory snatches of time," Dawson complained that "nothing has been perfected so far as it might."[2]

Dawson recounts that his final years in Nova Scotia and his early years in Montreal were spent conducting a "vast number of minor researches," dictated by specimens friends brought him or by inquiries at McGill's museum. The fruits of these investigations were published in the *Canadian Naturalist* and other journals, but after the passage of years these papers "almost escape[d] his own recollection." He recognized that his work as a teacher had compelled him to "spread himself" over a wide range of scientific subjects, and that much time had thereby been "frittered away."[3]

Dawson's investigations into the Precarboniferous fossil flora of Nova Scotia stand out as one of the few instances where he permitted himself to devote long periods of "continuous, sustained application" to a fairly narrow scientific speciality.[4] As early as 1861, he wrote Lyell about his fervent wish to produce a long treatise on these plants, characterized as one of his "favourite objects of study." Yet few British scientists recognized the importance of these fossils. As Dawson said to his confidant: "Your men who work in the field do not seem to know what they find, and your London botanists do not know how things are in the field."[5] His desire to publish a monograph on Devonian plants – a desire met by British ignorance of, and indifference to, these fossils – caused Dawson to endure "one of the deepest mortifications of my scientific life." Years later, he still viewed the whole affair with deep sadness, recollecting it as even more discouraging than his rejection by the University of Edinburgh.[6] The incident turned Dawson away from original research, directing his scientific fervour toward popularization and more doctrinaire matters.

THE BACKGROUND

It was a remarkable breach of tradition when the Royal Society of London refused to publish Dawson's Bakerian lecture of 1870, on the Precarboniferous flora of Northeastern America, in their *Philosophical Transactions*. Dawson's indignant and immediate response was to "throw no more of my pearls before the swine in that quarter." Even with the hindsight of twenty-five years, Dawson was to speak of his "great disappointment" over the shabby treatment of his "magnum opus" at the hands of the Royal Society of London.[7]

Dawson became a Fellow of the Royal Society in 1862, having managed to secure one of the few openings earmarked for a geologist.[8] His certificate of membership was blessed by several British geological "lions" of the day, including his patron and mentor Lyell, the Geological Survey of Canada director Roderick Murchison, and Murchison's successor, Andrew Crombie Ramsay.[9] Less successful was his bid in the late 1860s for a Royal Society of London grant to publish his work on the Devonian flora of Canada. (Not that Dawson's efforts were likely to have met with success: the society's limited endowments rarely funded publications or went to those far removed from London.[10]) The society appears to have offered Dawson the Bakerian lectureship as a consolation prize, fearing, perhaps, to call into question his patronage by Lyell and his ostensible support by Murchison. For his part, Dawson was justifiably proud of the prestigious award. At the time, the society had only one other lectureship, the Croonian, as well as three medals: the Copley, Royal, and Rumford.[11] The Bakerian lecture carried an endowed stipend as well, an impressive achievement for the colonial who often nattered about money matters, including the high cost of joining the Royal Society in the first place.[12] Moreover, since the Bakerian lecture had traditionally appeared later in the Society's *Philosophical Transactions,* often as the leading article, Dawson naturally assumed that publication there would follow.[13] (Humphry Davy's Bakerian lecture had even received a medal.)

Alas for Dawson, precedent was about to be broken. The Bakerian lecturers had indeed included a stellar group; numbered among them were J.W. Herschel, Charles Lyell, G.B. Airy, Richard Owen, William Thomson, and James Clerk Maxwell. Several, such as Humphry Davy, J.D. Forbes, Michael Faraday, Edward Sabine, and John Tyndall, had even won lectureships on more than one occasion and were still being rewarded with publication of their communications in the *Philosophical Transactions*. Dawson's lecture, however, appears to have ushered in a new phase of the Bakerian endowment: publication in the prestigious quarto journal did not inevitably follow. This new development did not, though, reflect on the quality of Dawson's work; subsequent lectures by Arthur Schuster (1884), J.J. Thomson (1887), and Norman Lockyer (1888) were also published only in the less prestigious *Proceedings* of the Society.

This gradual modification in the publishing pattern of the Bakerian lecture reflected a larger shift in the historical development of the Royal Society's official publications. Until about the mid-nineteenth century, communications to the Royal Society had been published in either of two publications: the *Proceedings,* which received shorter or less important works; or the *Philosophical Transactions,* which contained longer or more valuable papers. This distinction began to erode, however, as a number of significant papers began to appear in the *Proceedings*. At the same time, *Philosophical Transactions*' authors ceased to be regarded by the society as an elite group: the special reduction in their life membership fees was removed, and they were no longer distinguished in the fellowship lists.[14]

At this juncture, in subtle ways that Dawson could scarcely perceive and certainly not control, these new directions combined with established traditions to doom the ultimate publication of his lecture. Clearly, natural historians were "odd men out" in the galaxy of Bakerian lecturers, despite the fact that they were explicitly included by the terms of Baker's bequest. Only two others, Lyell (1835) and Owen (1844), appear in a roster otherwise dominated by physical scientists, save for a handful of physiologists. (Both the Copley and Royal medals were more frequently bestowed upon natural historians, whereas the Rumford Medal and Croonian Lecture were designated for other specialities). Furthermore, Lyell and Owen – like their fellow physical scientists – delivered technical accounts of research on restricted topics, not the kind of general survey that Dawson attempted to provide. For them, the Bakerian lecture was a communication like the many others delivered on countless occasions to the Geological or the Linnean societies; for Dawson, it was the summation of several decades of research and reading. By blowing the award out of proportion, he lessened the likelihood of its subsequent publication by the Royal Society.

Dawson's misunderstanding of the nature of the lecture was related to his relative isolation in Montreal, where he was not privy to the unarticulated protocols that guided British science in the metropolis. It was rare for a foreigner to win the Bakerian lectureship; it was unheard of for a colonial to walk away with the honour. A further complication for Dawson was that his sources of information on London science – Lyell and Bigsby – were providing unreliable reflections as to the motivations and interests of the emerging coterie of younger naturalists. Lyell, though "a public figure of venerable proportions," was seventy-three.[15] Bigsby, despite his active engagement in the affairs of the Geological Society, was nearly eighty and had only just become a Fellow of the Royal Society in 1869.[16]

As well, unbeknownst to Dawson, his timing was bad: a reluctant Royal Society had just agreed to publish another long paleobotanical treatise by Dawson's friend and colleague at Manchester, the medical doctor William C. Williamson. In the end, the nineteen installments of Williamson's work tied up the presses for more than twenty years.[17] It is hardly surprising, therefore, that the Royal Society would drag its feet when faced with more of the same.

THE LECTURE

For a while, the honour of delivering the Bakerian lecture – coinciding as it did with several other speaking engagements in London – appeared most auspicious for Dawson's scientific fortunes. At the end of April 1870, he was informed by the Royal Society's president, Edward Sabine, that the lecture would take place at the meeting of 5 May, a prime date on the social calendar of scientific London.[18] He appears to have been Lyell's houseguest for about a month, while being fêted, wined, and dined by leading scientific celebrities. He wrote to his daughter that he had met "no end of people," and sent her a newspaper clipping for the Montreal *Gazette* "that they may see a little of what I am doing here."[19] Sir Joseph Hooker offered Dawson the use of his house; in addition, Dawson managed to see Bigsby, Carpenter, William Carruthers, and Williamson.[20] The former director of the Geological Survey of Canada, Sir William Logan, lent him a map for his lecture and promised a new endowment for McGill University.[21] As well, Dawson, having carried a dozen large boxes of specimens with him to England, displayed some fossils at the Royal Society headquarters in Burlington House.[22]

Dawson's Friday night lecture at the Royal Institution (a popular version of the Royal Society lecture), which took place three weeks later, included plenty of pictures, a dozen or so specimens, and a few "striking facts."[23] Successful lecture topics had included Mt. Sinai, iron ships, and John Ruskin; most of the auditors were women.[24] Son George, then a student at the Royal School of Mines, warned his father not to expect a large audience, since only well-known "experimenters" attracted a crowd there. He also suggested that the lecture be limited to an hour, or the audience would leave, and urged "simplicity of title." He recommended using the term "vegetation" rather than "flora" and avoiding confusing terms such as "Laurentian Rocks." (Newspaper reports of the lecture show that Dawson heeded this advice.) Dawson was subsequently invited to display his specimens at a British Association soirée in Liverpool.[25]

Dawson's promising situation began to change later in the same month, when two fellows of the Royal Society, Hooker and Martin Duncan, were asked to judge the suitability of his paper for publication. As the secretary later explained, this kind of referral was standard procedure. From Dawson's perspective, however, the choice of Hooker was unfortunate, accomplished and esteemed though he was in all matters botanical. The two scientists had clashed eight years earlier when, as Hooker put it, Dawson had "poohpoohed his Greenland paper."[26] Hooker was unlikely to forget Dawson's negative assessment, nor to forgive the underlying anti-evolutionist position.[27]

Hooker, an outspoken Darwinist and a known "radical" in scientific and religious affairs, was bound to conflict with Dawson's pronounced conservatism in these realms.[28] But Hooker had climbed high on the social ladder of the scientific establishment; three years later he would be sworn in as president of the Royal Society. Devout but gossipy Bigsby complained to his protégé Dawson at Hooker's inauguration: "I do not at all like the hands into which the Royal Society and the Royal Institution has fallen – Belief in no God and no Bible is openly paraded."[29] These were the days when the agnostic and materialist X-Club – with Hooker at its helm – exerted a major influence over the affairs of the Royal Society, orienting it above all else to the advancement of science "untrammelled by religious dogmas."[30] In Dawson's view, Hooker (along with fellow X-clubbers T.H. Huxley and John Tyndall) stood among those "leaders of thought" whose religious skepticism was "eating the heart out of English geology."[31] Hooker's opinion, though, weighed mightily in scientific London, to the detriment of Dawson's scientific future.[32]

THE MANUSCRIPT AND ITS EVALUATION

The Bakerian manuscript that went out to Hooker, "On the pre-Carboniferous Floras of Northeastern America, with special reference to that of the Erian (Devonian) Period," began with three sections describing the fossil flora of the American Devonian formation. Dawson called the system "Erian" (following the New York Geological Survey's practice), because of its proximity to Lake Erie.[33] A fourth section, encompassing "general remarks and conclusions," was more likely to prove controversial given its consideration of the "Bearing of Erian Botany on Questions as to the Introduction and Extinction of Species." The entire discussion ran to more than 150 manuscript pages, clearly more than Dawson might have read before the Royal Society or, as it turned out, hoped to have them publish.

The heart of Dawson's lecture catalogued 121 species of fossil plants of the Eastern portion of this formation, characterized by *Prototaxites* and *Psilophyton,* which fell between the Upper Silurian and Lower Carboniferous. In the case of previously described species, he referred his readers to the appropriate memoirs. He added descriptions of twenty-three new species along with "corrections of errors, new facts and structures recently obtained, and discussion of the nature and affinities of the several species; so as to bring the whole subject as far as possible up to the present state of knowledge."[34] Throughout the memoir, Dawson insisted upon the importance of the Erian or Devonian period in the geological history of North America. Its deposits ran to a depth of 15,000 feet, covered nearly half the continent, and showed the introduction of a "rich and varied Flora," in addition to fishes, insects, and a "new series of marine forms." The continent should be seen, argued Dawson, as the typical region of the Devonian, and hence geologists everywhere should adopt the term Erian.[35]

Of course, Dawson's "Amerocentrist" attitude and nomenclature were not to win him supporters among European naturalists. In Western Europe, Dawson continued, the Devonian formation was "comparatively depauperated" in its fossil record, covered less breadth, and fell more subject to local variations – in short, it was imperfectly developed. Furthermore, argued Dawson, the most typical organism of the Erian, the genus *Psilophyton,* went undetected in Europe due to either defective preservation or insufficient observation. Dawson warned of the critical importance of studying fossil plants among the fossil beds themselves (rather than of observing specimens stored on museum shelves), where context and quantity compensated for individual imperfections. Because of their relative ignorance of, and disinterest in, Devonian fossils, Dawson expected naturalists in England to be skeptical of his work and his proclamation of the significance of these fossils elsewhere.[36]

Perhaps the content and tone of Dawson's argument perturbed his referees, unaccustomed as they were to seeing this genre of communication crowned by the Royal Society. Dawson himself noted at the end of his lecture that he might be criticized for presenting the general principles that guided his study of palaeozoic floras.[37] Indeed, most Bakerian lecturers simply communicated recent research, rather than delivering a lecture of general interest or a popularization. Lyell's lecture of 1835 – to choose a comparable example from the geological realm – had simply reported his observations of the previous summer.[38]

At this juncture, Dawson might have usefully recalled his mentor's description, nearly a decade before, of the publishing standards of the *Philosophical Transactions*. Lyell had cautioned that, above all else, a paper intended for that journal required careful construction and unhurried preparation. The manuscript of Dawson's Bakerian lecture, with its elaborate editing and last-minute scribbling, appeared to be neither. Dawson's lecture fell short on a number of other counts as well.[39] It was too long, and although the title was "severe", as suggested by Lyell (except for the somewhat bizarre nomenclature represented by the term "Erian"), the paper represented a simple summary of what was known and something "too popular and indefinite," rather than a new fact or theory, as required. Finally, the inherently controversial tenor of the paper was sure to offend the standards of Victorian science publishing.[40]

By mid-May, Hooker sent an eleven-page letter to George Stokes, one of the Royal Society's secretaries, detailing his objections to Dawson's paper.[41] Hooker explained that he had never before experienced such anxiety as a referee. On the one hand, Dawson's reputation as a geologist was established and high; on the other, said Hooker, "the descriptive part [of the lecture] contains little real novelty, and too many insufficiently characterized new genera and species of plants of unknown or altogether doubtful affinity." He took special issue with Dawson's use of fragmentary evidence – "a mere scrap of vegetable tissue, whether of root, stem, or branch" – as a means of classifying the fossil plants in question.

But it was when Dawson made far-reaching arguments about the origins and distribution of plants, based upon this evidence, that Hooker became enraged. Dawson's speculations were simply "astounding," said Hooker, when "we are ignorant of the actual origin of any one existing Flora on the globe, though we know in considerable detail the geographical, geological, and botanical characters of many such Floras." He complained that Dawson employed terminology in a loose way, using words such as "Archaic," "prototypic," and "synthetic," that "have no scientific value and convey no exact knowledge." Moreover, noted Hooker, the "loosely worded aphorisms" in the concluding section simply amounted to "principles familiar to every tyro in Natural History Science." Hooker contended that Dawson had not conducted a close enough examination of European type specimens, thereby nullifying the possibility of definitively linking them with American fossil plants. Furthermore, he argued that Dawson's illustrations were markedly inferior to those supplied in continental or British publications.

Any one of these comments would have been enough to suppress Dawson's paper; in their entirety, they demolished it. And although the official report was bad enough, Hooker sent an even harsher critique to his confidant Charles Darwin. Dawson, he said, had not contented himself "with the proper summation" of his discovery of certain Devonian plants; the result, wrote Hooker, was a paper filled with "perfect trash." He concluded by summarizing his recommendation not to publish the "disaster": "The systematic part is very meagre indeed, the vegetable anatomy miserable and often utterly wrong; the affinities more often mere guess work than not; and as to the theories and speculations, they would make your hair stand on end."[42]

The measured comments of the other referee, Martin Duncan, who held the chair of geology at King's College (London), contrasted sharply with Hooker's outrage.[43] Duncan explained that he was familiar with Dawson's work on Palaeozoic flora, and that the publication of this "very exhaustive" memoir "as it stands" would be useful and do credit to the *Philosophical Transactions*. But he, like Hooker, urged the excision of the chapter on the origin of species which appeared "quite out of place." Duncan concluded: "The rest of the memoir will last as long as science, but this particular chapter may be disowned by the author with much pleasure at any time within a few years."

The divergent assessments of the two readers lead us to scrutinize the refereeing procedures of the Royal Society of London more closely. First, neither Duncan nor Hooker was expert in paleobotany. (One wonders, for example, why Williamson was not consulted.) Hooker at that point in his career worked as a biogeographer, and he had not pursued fossil botany for at least fifteen years. He called it, in fact, "the most *unreliable* of sciences." Even a decade before, Dawson had complained to Lyell that Hooker seemed "misty and shallow" in his fossil botany, and "not well read up lately."[44] Duncan, at least, was an expert on fossil corals.[45] Both, however, were thoroughgoing evolutionists, if not Darwinists, which could imply – as it did in the case of Hooker – a strong ideological opposition to whatever Dawson wrote. By this time, London scientific thought had shifted to the side of the evolutionists (Darwin had received the Royal Society's most distinguished award, the Copley Medal, in 1864), and the Canadian critic of evolution was being left on the sidelines.

Despite Dawson's network of correspondents who regularly forwarded news of things British, he had no inkling of the storm brewing over his lecture. In June he deposited a letter and memorandum with the Royal Society, expressing his concern that the lecture be published in its entirety, fully illustrated, and offering to pay, should this be required (clearly his finances were more elastic than he had earlier indicated).[46] At the beginning of July, Walter White, the assistant secretary, simply informed Dawson that the Committee on Papers had not yet reached a decision about his memoir, even though he had received Hooker's report months before.[47] In the long weeks that followed, Dawson must have become agitated when no new word arrived from London. By early November, he wrote to Lyell, asking him to check on the progress of his paper. He would prefer to publish it in Canada, he added, rather than face further delays.[48]

Secretary Stokes had a difficult decision to make: he had two emphatically different evaluations of Dawson's lecture. He dithered six months, before soliciting a third opinion, this time from William Carruthers – another "establishment" figure, soon to become keeper of botany at the British Museum. Although Carruthers had told Dawson in an earlier letter that same year that "the Devonian Flora is so completely your own,"[49] he supported Hooker's negative assessment – not surprisingly, having seen Hooker's long criticism as well as Duncan's report.

Again, fate intervened in the form of bad timing. Just at this point, Dawson himself was taking issue with Carruthers's work, particularly his identification of a Devonian *Nematophyton*. Bigsby was later to characterize Carruthers as "positive and tenacious of his opinions botanical – and probably combative," as well as "unpleasant in printed disputes."[50] Referring to Carruthers, Dawson complained to Lyell about "all the mistakes that his too technical botany ... have led to," and proposed, once again, that he study "plants as they stand in the cliffs at Sydney and the Joggins, instead of on the shelves of the British Museum." He concluded with a note of despair:

However well and carefully work may be done here, no value is attached to it in England. I fear it will soon come to this that both in chemical geology and fossils we shall in America have to cut all connection with England; all your work is becoming so thoroughly shallow and imperfect insofar as things not of Europe are concerned, and we are too far off to correct it or even to defend ourselves.[51]

Dawson, whose paleobotanical investigations would be acclaimed for their delineations of elusive relationships between organisms, naturally took issue with a narrow approach that advocated interpreting specimens outside the matrix of their paleoecology.[52]

During the early part of 1871, Dawson continued to express anxiety about the fate of his paper to Lyell and Bigsby.[53] He wanted to find out who was refereeing the paper. Lyell responded that he was not on the society's council and "it might be thought very indiscreet in me if I were to enquire."[54] Bigsby wrote that he was not in a position to ask the names of the referees and admonished Dawson that "it is better for you not to know."[55] He later expressed his puzzlement over the whole affair and attempted, without much success, to piece together what had happened.[56] Although Bigsby later admitted to having laid, in vain, "some honest traps to find out how it came to pass," he urged Dawson "not to fall into personalities."[57]

Only the following spring, a year after the lecture, did Dawson learn, much to his surprise, that the cost entailed by the number of plates was not at issue, but that, indeed, the society had rejected his memoir altogether. Stokes explained that Dawson's eminent scientific position forced him to provide the reasons for the negative decision, rather than simply sending him a formal announcement. He summarized the reasons as follows: the committee objected to the "far too slender evidence" supplied by Dawson and had scolded that "doubtless the paleontologist in the eager pursuit of his favourite science is tempted to make the most of his scanty material; and ... to supplement what he has got by conjecture." Publishing such "precarious" arguments in the pages of the *Philosophical Transactions* would "run the risk of serving to stereotype error."[58]

To Dawson, these strong words merely reflected the "low state of paleobotany" in England. His chief concern at this point was to retrieve the original manuscript and drawings, in order to publish the memoir in Canada (presumably the manuscript was officially the property of the Royal Society, as was the case once a paper was read to the Geological Society).[59] He explained that the delay had already been considerable, that colleagues were awaiting its publication, and that the original copy contained many notes and additions not included in his version.[60] Dawson asked that the whole be turned over to Lyell, and even offered to return the fee received for delivering the Bakerian lecture.

The Royal Society Council promptly sent back Dawson's drawings and his copy of the manuscript, which was subsequently published, in part, by the Geological Survey of Canada.[61] He wrote to one member that he would send the Royal Society of London no more papers, but that if "any of the small men they regard as botanical authorities in London should presume to criticize my paper, now that it has been printed by an institution more liberal and enlightened than the Council of the Royal Society, I shall retaliate in such a manner as will bring before the public the whole subject."[62] Dawson distributed a printed circular to council members which explained why he had published his lecture elsewhere; he warned that he would resent any further criticism as "impertinent and offensive."[63] He believed that his work should have been seen as a "godsend" in remedying the Royal Society's "utterly below par" competence in paleobotany.[64]

Privately, Dawson complained to Lyell somewhat less bombastically, remarking on the poor work of English botanists and urging the attendance of "your botanical men from Hooker downwards through a course of [the American botanist] Asa Gray or some similar elementary work."[65] Perhaps to console himself for his work's issuance as a report in local geology, he commented that far greater influence would be obtained by publication in North America, and those few individuals in England interested in the topic could easily procure copies. Finally, he noted that unless he found some way to retire to Britain in the near future, "I may as well give up the idea of pursuing geological subjects except with reference to the American public."[66]

THE MORAL

The story of Dawson's troubled relationship with the Royal Society continued through several more decades, but it is interesting to speculate why the Bakerian lectureship incident worked out the way it did. Clearly, Dawson's pronounced opposition to evolution antagonized scientific powerbrokers like Hooker. His geographically peripheral position served, moreover, to enhance his ideological marginality. Both problems were exacerbated by imperfect and incomplete communications with the metropolis, Dawson being dependent on the perceptions of two aging scientists, Lyell and Bigsby, who seem to have had less than perfect comprehension of the changing motivations and concerns of London scientific luminaries.

But the variety of science – the particular subdiscipline, paleobotany – that he had chosen to cultivate would seem to have further ensured his marginality and allowed the Royal Society to dismiss his important work. There seems to be no question as to the quality of his contributions to this field; one American paleontologist of his day called him "the recognized authority on the Devonian flora on this side of the Atlantic."[67] Even today he is acclaimed for his "shrewd and critical judgment," and his "wide and imaginative research," despite the fragmentary specimens with which he worked.[68] He described 125 new species of Palaeozoic plants (including twenty-two Devonian) that still stand today. One paleontologist has commented: "The microscopic detail he was able to wring from even poorly prepared specimens should fill today's experts with envy ... it is surprising not that he did so well, but that he was able to accomplish anything worthwhile."[69]

One is tempted to read into the choice of such referees as the biogeographer Hooker, the coral specialist Duncan, and the botanist Carruthers, a deliberate attempt to scuttle Dawson's work in an unusual subdiscipline or interdiscipline, fossil botany. Yet as T.G. Vallance points out, no British university, museum, or government agency employed a paleobotanist as such. The *Times* called paleobotany "a neglected science" in England, in contrast to Europe and North America where its relevance to coal geology and mining was better understood. In addition, argues Vallance from the Australian case, European scientists were reluctant to acknowledge the existence of fossil organisms that did not fit into their notions of lithological stratigraphy.[70] Because of the lack of a genuine interest in training and supporting those who worked to decipher the fossil record, diverse paleontological opinions and interpretations were fighting for supremacy in London at this time.[71]

Interestingly, Dawson, for his part, continued to bear grudges against individuals such as Hooker, whose hand, he correctly deduced, showed in the affair. Years later, Dawson wrote Hooker that it was a "disgrace" to "English science" that he had promoted European paleobotany, but "slighted" what had transpired in Canada.[72] Even more important for the future direction of Dawson's own scientific research, he viewed the incident as more than a "slight" – it constituted a source of rejection, mortification, and even depression. For a long time afterwards, Dawson took little interest in scientific investigations of any kind. Not only had the response to his Bakerian lecture proved to be his "bitterest lesson as to the variety of trust in man," it "threatened to be a death blow" to his original scientific work. Once the scars began to heal, Dawson returned to the world of science, but with a fundamentally altered perspective toward his work. Specialized and abstruse pursuits no longer suited his scientific temperament. Henceforth, he determined to make his mark as a popularizer and generalist.

9 A Mission of Popularization

The poor reception to Dawson's work on Devonian plants, although it naturally diminished his enthusiasm for this line of research, served to rekindle his interest in the popularization of science. He later viewed the affair as an important lesson that caused him to take a stand against "the false philosophies of the day." In retrospect, he was to see his increasingly strong commitment to popularization over the years as perhaps having "crippled my scientific reputation." But in his opinion, it possessed the more important benefit of compelling him "to do much for the moral and spiritual good of man."[1]

Certainly, the events of the early 1870s renewed Dawson's resolve to popularize science, which often, in his view, meant relating scientific advances to the teachings of scripture. But Dawson's interest in popularization also can be understood as a natural extension of his grappling with intellectual issues as a young man not yet thirty. Foremost among his concerns was a desire to harmonize his scientific pursuits with his religious interests – a legacy, perhaps, of his youthful days at Reverend Thomas McCulloch's Pictou Academy, where scientific and religious enlightenment worked in tandem.[2] In particular, Dawson sought to demonstrate that the reflection of God's handiwork displayed in the geological record accorded perfectly with divine wisdom as expressed in the Bible. He explained to Lyell at the time that he was "attacking" Hebrew and assembling all available works of biblical criticism, in order "to study the view of natural history given in the Bible." He concluded that this view was "much more precise and systematic than usually supposed," and that translators and commentators had failed to appreciate its importance because of their scientific ignorance. Although Lyell probably found this a "hopeless subject," Dawson placed "equally strong faith" in "the inspiration of the scriptures" and "the results of accurate observation."[3]

Lyell warned his young protégé that Pictou had few resources to support his mission; indeed, he would have to study Sanskrit and other oriental languages in the libraries and under the tutelage of scholars in London, Oxford, Berlin, or Paris. Lyell argued that knowledge of God and the love of truth would be better advanced in Dawson's case if he were to devote his attention to exploring the wonderful natural formations around Pictou.[4] Lyell commented that he himself had "never enjoyed the reading of a marvelous chapter of the big volume more" than during their excursions to the remarkable cliffs at the Joggins.[5] But Lyell's words failed to impress the young Dawson, who at that point foresaw a career as a minister, not a geologist. He even described one of his early scientific publications, on the *Geography and Natural History of Nova Scotia,* as adopting the convenient format of William Pinnock's *Catechisms.*[6] It is hardly surprising that Dawson, so immersed in paleontological and geological fieldwork and so imbued with the zeal of a self-confident Presbyterianism, should strive to reconcile these two aspects (almost two halves) of his personal existence. But his turning of this emotional quest into an intellectual odyssey would, in the end, prove costly to his scientific reputation.

During the years that followed, Dawson resorted to teaching himself, having failed to find formal structures for learning Hebrew. This experience in no way lessened his desire to reconcile Scripture with science, which Lyell called his "six-day system of geology." The elder geologist did, however, worry that Dawson was placing his mind in a straitjacket: he urged him not to let these views "prevent your expansion in theoretical power." Lyell even admitted that he was frequently accused of forcing new facts to fit the framework of his twenty-five-year-old *Principles,* but that he practised letting "out my clothes now and then when I feel them growing tight and stunting my growth." Lyell warned: If geology has changed so much in twenty-five years, just imagine the changes wrought in all departments of knowledge over twenty-five centuries, and the difficulty in making present understanding fit past conceptions.[7]

Lyell's words would turn out to be prophetic in terms of the great transformation about to occur in Biblical scholarship. Under the leadership of the Göttingen scholar Julius Wellhausen (1844-1918), the Old Testament would soon serve as the focus of critical, academic scrutiny according to principles established by historian Jakob Burckhardt. Wellhausen would conclude that the books of the Pentateuch had evolved from oral traditions, and did not represent divine revelation as conveyed by Moses.[8] Although Dawson wrote *Archaia* on the eve of this revolution, he scarcely acknowledged the shift in the analysis of the Bible brought about by German academics.[9] As a result, his critics blamed him for advocating the traditions of "old Presbyterian theology," fundamentalism, and evangelicalism.[10]

ARCHAIA

In 1860, Dawson published *Archaia; or, Studies of the Cosmogony and Natural History of the Hebrew Scriptures*.[11] Subsequent events would show that the title was simply too arcane to attract many buyers; finding more comprehensible terminology appears to have been a precondition to reissuing the work in the late 1870s.[12] Moreover, the title's resemblance to Dawson's earlier work *Acadian Geology* created needless confusion.[13] The history of *Archaia* goes back to 1855, when an advertisement in Scottish newspapers for the Burnett prize (and the lack of publishing opportunities in Pictou) apparently spurred Dawson to submit the manuscript. This competition awarded £1800 for the best essay "in proof of the existence of a supreme Creator, upon grounds both of reason and revelation." The young man from Nova Scotia failed to capture the purse (even the future Archbishop of Canterbury, John Bird Sumner, had been unsuccessful in a previous competition, forty years earlier.)[14] The essay failed to interest a publisher until five years after Dawson was inducted as principal of McGill, when the book was published simultaneously in Montreal (B. Dawson & Son) and London (Sampson Low, Son & Co.).

In his autobiographical *Fifty Years of Work in Canada*, Dawson treats *Archaia* as merely another link in the chain of his research and writing. Indeed, the geological record served as the starting point for both *Acadian Geology* and *Archaia*. Dawson explained to Lyell that his study of Hebrew cosmogony aimed to show "that the Bible ... encourages geological inquiry, and opposed none of its legitimate conclusions."[15] He bestowed lavish praise on American geologist James Dwight Dana's series of articles in the *Bibliotheca Sacra,* which established priority for modern science against the claims of the bibliolaters.[16] The young Dawson was not one to sacrifice science on the altar of religion. In *Archaia,* he cast a wider net than ever before, as he scrutinized a different subject matter with the methodological tools he had employed in the geological realm. He sought to apply the fruits of modern science to scripture in two mutually reinforcing ways.

First, he adopted the inductive method of William Whewell (whose words grace the title page of *Archaia*) to study the Bible. He strove to analyse the component "facts" with precision – that is, to understand the exact meaning of key terms in the original Hebrew. In the words of his mentor Lyell, his task consisted of "giving the internal evidence the best attention he can, and exercising his reasoning powers upon it to the best of his ability."[17] For Dawson, the divine status of the Bible was not challenged by being subjected to human interrogation. (He argued, in fact, that monotheistic religion encouraged all forms of scientific endeavour.) He felt that once he had cleansed the account of creation of its misleading connotations, for example, it could not fail to accord better with the discoveries of modern paleontology and geology.

This latter stage, then, represented the second way in which modern science could enlighten the reading of Scripture: an understanding of recent scientific accomplishments, such as La Place's nebular hypothesis, could be used to illuminate events recounted in the Bible. What was remarkable, in Dawson's view, was the tremendous influence of the Hebrew Scriptures on all subsequent theologies and philosophies; as a result, he had to "assign an important scientific place to all they say of nature." The importance of these references to nature, however, had been obscured by the "half light" shed by the "common run of expositors." In place of these "dark gropings," Dawson insisted that it was necessary to examine Scripture itself, brushing aside the theological controversies of the day for its revelations of past and future.[18]

Dawson explained in the preface to *Archaia* that the work was the "result of a series of exegetical studies of the first chapter of Genesis, in connection with the numerous incidental references to nature and creation in other parts of the Holy Scriptures." He proceeded by discussing, in turn, the meaning of words like "created" (*bara* in Hebrew), "day" (*yom*), and "herbaceous plants" (*deshé*) in order to present an original and positive interpretation of Scripture as it related to modern geology. This involved a rejection of the synthesis of the American geologist and minister Edward Hitchcock (in his *Religion of Geology* of 1851) and a modification of the views of the Scottish pastor Hugh Miller (in *The Testimony of the Rocks* of 1857 and elsewhere). Not only did Dawson view their interpretations as misleading; he saw their works as carrying a more serious flaw: in attempting to "reconcile" science and religion, the authors had adopted a defensive posture and settled for presenting their views within a negative framework.[19]

Dawson argued in *Archaia* that the Mosaic cosmogony, as described in Genesis, must be accepted as a direct revelation from the Creator. Given its special status as "inspired teaching," attempts like Hitchcock's were ill-conceived glosses on the gospel, which by modifying the

Divine Word served to minimize and misconstrue its import. Nonetheless, his reverence for the Word did not impede Dawson from attempting to ascertain the correct connotation of Hebrew terms, whose meaning he believed had been adulterated over the centuries. Much of his analysis hinged on the meaning of a day of creation, which he insisted was not a natural or civil day, but an indefinitely long period of time.[20] Dawson, in other words, insisted on the divine nature of Scripture, but believed that it was the responsibility of scholars to employ scientific tools, whether methodology or information, to tease out its proper meanings and implications.

Dawson endorsed the view that the fields of revelation and natural science were distinct and independent. He saw his work as blazing a pioneering trail that illuminated some of the "meeting points," as he called them, "of Biblical lore and geological exploration." Unlike those who minimized or impugned the validity of either tradition, Dawson saw as his unique role the pointing out of "certain manifest and remarkable correspondences between these teachings [of science] and those of revelation."[21] Although both modes of understanding might be expected to arrive "at some of the same truths, though in very different ways," it was sheer folly to expect them to agree "fully and manifestly."[22]

Archaia represents both a natural extension of, and a new direction in, Dawson's intellectual and literary productivity. It brought him into close contact with a more heterogenous group of correspondents, many of whom were better versed in the intricacies of biblical exegesis than in the arcana of scientific method and practice. For Dawson himself, the work launched his career as an expositor or popularizer of science. No longer did he have to scratch out a living or build a reputation from field work alone. The publication of *Archaia* also initiated Dawson's long and close relationship with the publishing firm of Dawson Brothers (no relation) in Montreal, who became known for their scholarly and scientific editions. By 1860, Canadian publishers were entering their most prosperous decade of the century, and Dawson's enormous productivity became part of this literary flowering. Dawson Brothers followed a typical Canadian pattern of building a reputation by publishing local authors (as well as local editions of British and American works).[23] In Dawson's case, their firm generally provided a Montreal imprimatur for works that were simultaneously issued in New York and London. *Archaia* found few buyers, but in a revised form as *Origin of the World* it went through six editions (by 1893). Similar and subsequent works were even more successful; *The Story of Earth and Man*, for example, saw eleven legitimate and several pirated editions.[24]

Dawson began to look to new audiences with the publication of *Archaia*. The book served to answer questions posed to him as "a teacher of Geology," but also aimed to provide geologists with "a digest of the cosmical doctrines to be found in the Hebrew Scriptures, when treated strictly according to the methods of interpretation proper to such documents, but with the actual state of geological science full in view." Furthermore, for "biblical students and Christians generally," *Archaia* showed how "the scriptural cosmogony presents itself to a working naturalist, regarding it from the stand-point afforded by the mass of facts and principles accumulated by modern science."[25]

Reviews of *Archaia* concurred that Dawson had well served both scientists and theologians. In the United States, the *Bibliotheca Sacra* thought that Dawson had fairly weighed the claims of nature and revelation, and pronounced his views "broad and Christian."[26] The *American Journal of Science* placed a short review of *Archaia* just before its first notice of the *Origin of Species,* and proclaimed that Dawson "like all devout and earnest men of science ... does not for a moment doubt that the genesis of the rocks will confirm the Genesis of Moses."[27] In Canada, as well, the *Canadian Journal of Industry, Science, and Art* mentioned Dawson's "sincere and strong faith in the divine truths of Revelation," and accordingly regarded *Archaia* as a contribution to the literature of the Bible. But the reviewer, Edward J. Chapman (professor of geology at University of Toronto), commended the work for "those who still blindly look upon geology, and upon natural science generally, as antagonistic in some undefined manner to the spirit of Revelation."[28] And the Reverend Alexander Kemp, a long-time friend and mentor of Dawson, devoted more than twenty pages to the book in the *Canadian Naturalist*. Kemp proclaimed the book as "the beginning of a new period in the Literary History of Canada." Here a "colonial author" had chosen not to treat local issues for indulgent readers, but opted for "embracing fields of investigation of universal interest" and for "challenging the attention of both religion and science."[29] Accordingly, the author had left the common path of fact gathering to scale the dizzying heights of theoretical construction and speculation.

Dawson – who saw his mission as that of conciliating religious and scientific traditions – grew increasingly impatient with what he saw as excessive zeal or short-sightedness on either part. On the one side were the "bigots"; on the other, the "infidels."[30] He explained to Lyell (who had earlier expressed his "despair" at being unable to reconcile modern geology with ancient cosmogonies) that the account of creation in Genesis provided "one of the best antidotes to the pantheism and rationalo-ritualism which are eating the life out of both science and religion in England, and which the flood of trashy popular science and of theology washes even to us here."³¹ A letter from the Oxford professor and medical doctor, Henry Acland, expressing his dual allegiance to science and religion, echoes Dawson's sentiments precisely, except that for Dawson there could be no question of the existence of God:

I am willing to be put in Coventry by both [science and religion]. But I will not in this matter for peace sake violate my conscience nor cheat my reasoning faculty. It is not in the nature of things that true religion and true science should be parted. If there be a God, he is Lord of my body as of my mind and equally. And the laws of the material and of my spiritual being must be alike His – if He is.³²

ORIGIN OF THE WORLD

By the time the second edition of *Archaia* appeared in 1877 – newly entitled the *Origin of the World, according to Revelation and Science* – Dawson could identify himself as the author of *Acadian Geology*, as well as *The Story of the Earth and Man* (1872), *Life's Dawn on Earth* (1875) and *Nature and the Bible* (1875). With the re-edition of *Archaia*, Dawson associated himself ever more strongly with a tradition of dissent from academic science and affiliation with a religiously inclined audience. He was aware and concerned that these works would be "utterly condemned by the so-called 'leaders of thought' in London," realizing that a commitment to popularization would tend to diminish his reputation among scientists.³³ He asked geologist Sir William Logan, for example, not to be too critical of his *Story of Earth and Man* since it had been written under many difficulties and hindrances, and directed at a wide audience.³⁴

The new title of *Archaia*, the *Origin of the World* – chosen to more clearly indicate its character and purpose³⁵ – seems almost deliberately geared to provoking discussion with its allusion to Charles Darwin's *Origin of Species*. It reached, however, diametrically opposing conclusions. And substituting for the earlier quotation from Whewell (either incomprehensible or irrelevant to the book's readers) appeared the admonition from Job, "Speak to the Earth, and it shall teach thee." These changes made the volume more popular, for the *Origin of the World* was to sell nearly 1500 copies during its first year.³⁶

By this time Dawson had ceased to be satisfied with merely attempting to reconcile science and religion, and had begun to adopt a new course as a defendant of Christian theology. Instrumental in this transformation was the dissemination of Darwin's theory of evolution.

In Dawson's opinion, the Darwinians had committed countless "absurdities."[37] As his religious convictions increasingly removed him from the arena of scientific debate, Dawson found himself amid those who worried over theological interpretations. As he stated in the preface to the *Origin of the World*, he intended "to throw as much light as possible on the present condition of the much-agitated questions respecting the origin of the world and its inhabitants." But, because Darwin's theory had renewed the antagonism between science and religion, placing revelation on the defensive, Dawson believed it his personal mission to redress the imbalance.

In the view of Dawson – an individual who believed fervently in the union of the word and the work – a conflict had arisen only because one or the other tradition had lost sight of its true nature. The *Origin of the World* was intended to advance a return to original meanings and functions, given its aim "to rescue science from a dry and barren infidelity, and religion from mere fruitless sentiment or enfeebling superstition." Even science, Dawson argued, had been in the process of erecting "a barrier of scientific fact and induction" against the "crude and rash hypothesis" of evolution, by drawing upon the nebular hypothesis, the correlation and conservation of forces, new estimates of the age of the earth, the overthrow of the doctrine of spontaneous generation, philology, and anthropology. Adoption of these new theories helped to cleanse the Bible of "those accretions of obsolete philosophy" and revealed its "actual views of nature."[38] But following the publication of Darwin's *Origin of Species*, the scientific tradition appeared to have adopted an aggressive, militant stance. Dawson's writings sought to curb this posture, however outspoken he might appear to his opponents.

Nor were Dawson's views extreme, compared with those of some of his correspondents who railed against the Darwinian "infidels." J.J. Bigsby in England foresaw a "great fight" between the friends of Christ and the men of science. It was unfortunate, in Bigsby's view, that individuals such as John Tyndall and Thomas Henry Huxley were misled by "evolution à la Darwin," when at a personal level they did so much good. He painted a bleak picture of a growing social malaise, where "all classes of society in England are in various extremes, with wide deserts between them." Even amid the sanctuary of the Geological Society meetings, he was "shocked at the indifference with which the infidel opinions" of some members were received.[39]

A common clerical response to the *Origin of the World* was that such works provided a useful antidote not so much to Darwinian evolution as to rampant scientific materialism. E.J. Craven, director of the Princeton Theological Seminary, complained of the pernicious influence of "such men as Huxley, Tyndall, [John William] Draper, etc." with their "unchallenged mis-appropriation of the terms science and scientists." As "an acknowledged master in Physical Inductive Science," Dawson's opinion counted on these issues.[40] The British physician Cuthbert Collingwood saw Dawson's book as instrumental in suppressing doctrines of such men as John Fiske, Herbert Spencer, and "all the tribe of Nature-worshippers."[41] In India, the American missionary James Smith also praised Dawson's book for countering the widespread notion among educated young Indians that "Genesis and Science are inconsistent and that Science has forced the Bible to the wall."[42] Reverend J.C. McCulloch wrote from Australia that he found the *Origin of the World* to be the best book he had ever read on the subject, most other such works having been written by "theologic men ignorant of science, or by scientific men ignorant of and opposed to religion."[43] Even at home in Montreal, the rabbi and McGill professor of Hebrew, Abraham De Sola, thought the volume would strengthen the hand of those fighting against the "furious and unfair onslaughts of the enemies of the Hebrew Scriptures."[44]

DAWSON AS POPULARIZER

Dawson scarcely modified the views expressed in *Archaia* in his subsequent articles and books that addressed the same theme of reconciling science and the Bible, his bias becoming ever more pronounced on the side of Scripture. Indeed, lengthy additions and extensions were the rule.[45] Dawson's increasingly prolific writing on the topic actually reflected the surge in invitations to lecture from theological seminaries and church groups. During the mid-1870s, for example, he was asked to deliver a series of six lectures to the Union Theological Seminary in New York City on "the relation of the Bible to the sciences." There Dawson argued that unless naturalists were free to pursue their work, "untrammeled by the methods of theology," their results would be valueless. Dawson brought to the task of reconciling science and religion introspective "knowledge of the closet," as well as knowledge gleaned from "the field, the forest, and the mine." Yet the published version of these lectures, entitled *Nature and the Bible*, was seen as a better defence of religion than science. In the words of the reviewer for the *Canadian Naturalist*, the work formed "a handy repertory of replies to many of the current attacks upon the Christian theory of the system of nature."[46] Dawson argued that the sophisticated knowledge of the natural world displayed in the Pentateuch established that Moses had possessed special sources of information. It was irrelevant, in Dawson's view, how and why Moses had obtained this information, or indeed whether he was, himself, the sole author of the books.[47] (Thus Dawson dismissed the tradition of German scholarship that was beginning to transform knowledge of the Old Testament.)

Periodicals such as the Religious Tract Society's *Leisure Hour* (and its cousin across the ocean, *The Princeton Review*) also solicited articles from Dawson. Dawson was paid handsomely for these popularizing activities – not uncommonly he received hundreds of dollars for a series of lectures or articles. The editor of the *Leisure Hour*, apparently acting on the advice of Bigsby, promised Dawson "the highest rate" that the journal could afford.[48] The president of Drew Theological Seminary in Madison, New Jersey, assured Dawson that he could not begin to pay him what his lectures were actually worth, but would try to meet his price.[49] Dawson, however, managed to make his addresses and essays serve double duty by publishing them as books. The *Story of Earth and Man* (1873), for example, brought together articles that had been published previously in the Religious Tract Society's *Leisure Hour* during 1871 and 1872.

If the faithful rewarded and applauded Dawson for his literary accomplishments, they needed him for his scientific stature as well. J.R. Pattison, editor for the Religious Tract Society, admitted that "without your name the Committee would hardly venture on the line of argument or even on the Statement of facts given in your papers." He contrasted Dawson's publications to "the usual number of abortive reconciliation schemes appearing, as they do, every season promoted by the ignoramuses on one side or the other."[50] He maintained that there was nothing better or even equal to Dawson's contributions in the scientific domain, and nothing even remotely similar among the ranks of the theological.[51] Indeed, Dawson's contributions to their penny magazine, the *Leisure Hour* – which billed itself as "A Family Journal of Instruction and Recreation" – must have challenged a readership more accustomed to hearing about the "gardener's friends and foes" or "the great bustard" for their scientific edification. In his series of sixteen "Sketches of Geologic Periods as they appear in 1871," Dawson discussed the debate surrounding the nature of *Eozoön*, stressed the significance of the Devonian fossil flora, and even included a table taken from one of Bigsby's works. Dawson did adopt a tone of exposition for elucidating these obscure topics. He also provided magnificent and copious wood engravings for each of the geological periods.

Jonas Marsh Libbey, editor of the *Princeton Review* (one of the great literary quarterlies, unlike the *Leisure Hour*) encouraged submissions from Dawson as well. Libbey viewed Dawson's articles as a great asset inasmuch as he was someone "whose scientific researches have been so profound and carried on in so truth-loving a spirit."[52] Presbyterian clergyman Alexander Kemp equated Dawson's contributions with those by James Dwight Dana, James Hall, and the Prague paleontologist Joachim Barrande, saying that a single page from any one of them was worth volumes written by presumptuous theologians.[53]

Although Dawson brought his scientific knowledge to bear in these publications, it was the editors who schooled him in the art of popular expression. James MacAuley, an editor with the Religious Tract Society, urged Dawson to emulate Michael Faraday's style – also exemplified in Hugh Miller's writings – that "attract[ed] popular attention without lowering scientific tone."[54] Dawson even perused some of the works by the French popularizer Louis Figuier. He found Figuier's illustrations inaccurate, however, so he prepared his own drawings. Dawson hoped that his articles would be intelligible to the layman, yet still appear fresh and novel to the scientist, by "grouping the leading facts in a simple manner, yet avoiding what is too trite and commonplace, and bring[ing] out interesting deductions and relations to modern times."[55]

As for Darwin's theory of evolution, the editor of *Leisure Hour* asked for a simple "statement of facts and instances not to be explained by that theory," rather than any "declamatory talk," general arguments, or reflections, which would go "unheeded."[56] Similarly, the question of "primitive man" needed a "plain clear 'reading up' and exposition." Editor MacAuley remembered a passage by Hugh Miller which demolished Lamarck's hypothesis and might well be applied to help debunk later transformist theories.[57]

The editor of an American encyclopedia, impressed that Dawson's series of geological articles in the *Leisure Hour* revealed "mastery of the subject" and a "reasonable and candid spirit," asked Dawson to write the general entry on "geology." He offered him six dollars a page, as well as the opportunity to condense his *Leisure Hour* articles therein, even though no other contributor received more than four dollars a page.[58] Although Dawson himself worried about how scientists might view his popularizations – he did not wish "to be confounded with the numerous Reverends and other gentlemen who have taken up geology as amateurs"[59] – the book publishers, who always kept a careful watch on the balance sheet, urged Dawson to play down his scientific expertise. Hodder & Stoughton, for example, insisted that the word geology appear (if at all) in the subtitle rather than the title of a work on primitive man, in order to attract general readers.[60]

DAWSON AS ANTI-EVOLUTIONIST

Dawson's active denunciation of Darwinian evolution intensified his commitment to Scripture over science. He rigidly maintained this stance for forty years, from the publication of the *Origin of Species* in

1859 until his death in 1899, earning him the title of successor to the comparative zoologist and creationist Louis Agassiz. Early in 1860, he sent Lyell a copy of his book review of Darwin's *Origin of Species* (presumably written for the *Canadian Naturalist*), as well as some of his own insights into Darwin's difficulties. Although Dawson termed Darwin's research on variation good, providing "a wholesome check to species-makers, especially on this side [of] the Atlantic," he labelled his thoughts about species as "dreadful," amounting to a caricature of Lyell's "beautiful logic." Dawson feared that Darwin's book would do "much harm," its "tempting generalization" swaying many good naturalists, including Huxley and Joseph Dalton Hooker. He felt that most naturalists of the day were mere "collectors of facts," unlikely to recognize the errors in Darwin's argument. This deficiency, he noted, did not stem from "any defect in the facts of the case," but in the "coupling together two classes of phenomena which have no logical connection, and assigning as common causes terms which mean one thing when applied to one class of facts and another when applied to the other."[61] In other words, Dawson – like many of Darwin's critics – objected to a too-facile association of artificial and natural conditions.

Dawson's review in the *Canadian Naturalist* was less scholastic but no less negative. He stated his objections there in twenty pages, punctuated by long extracts from Darwin's book. Everywhere that he disputed Darwin's argument, he blamed a false methodology being used to replace the certainty of induction. In Dawson's view, Darwin had mistakenly linked the "proved ground of specific variability" to the "mystery of specific difference" in a "wild and fanciful application" based on "mere analogy." (Here again, Dawson's position was shared by skeptics who could not imagine how individual variability might give rise to interspecies differences.) Dawson argued that Darwin's belief in natural selection producing varieties and incipient species showed a "huge hiatus" in his reasoning.[62]

A common thread in this and Dawson's later writings was his conviction that Darwin had committed grievous methodological mistakes. Dawson differed from other critics only in the tenacity with which he upheld this view, unlike most naturalists who relatively quickly accepted Darwin's analysis. Dawson maintained that Darwin had constructed a misleading analogy between artificial and natural realms, claiming, further, that the hard empirical evidence of transmutation was missing: at best, one might "build up an imaginary series of stages, on the principle of natural selection, whereby these results might be effected; but the hypothesis would be destitute of any support from fact."[63] Acceptance of Darwinism, argued Dawson, meant that "our old Baconian mode of viewing nature will be quite reversed." As a consequence, he contended, "Instead of studying facts in order to arrive at general principles, we shall return to the mediaeval plan of setting up dogmas based on authority only, or on metaphysical considerations of the most flimsy character, and forcibly twisting nature into conformity with their requirements."[64]

In Dawson's view, the leap of faith required to accept Darwin's argument (based on "showy analogies") was matched by the ingenuousness necessary to believe that an imperfect geological record supported his claims.[65] The paleontology of the day, argued Dawson, provided no reason to favour evolution by natural selection over any other explanation of why the natural world appeared as it did. Rather than variation and change, Dawson espied constancy over time, in both fossil and modern forms. He cited examples from the shells and plants of the Post-Pliocene strata of Canada, as well as the Trilobites as described by Joachim Barrande. In other instances, he contended, old forms had become extinct and new types were introduced as modern conditions became too dissimilar from those of ancient times.[66] He fretted, however, that these facts were unattractive compared to the bold speculations of those "who can launch their opinions from the vantage ground of London journals."[67] (Dawson was not one to let his critics forget that he had been effectively disbarred from the most influential scientific journal of the day, and forced to turn to Canadian and popular journals in order to disseminate his anti-evolutionary message.)

In conclusion, given the faults in Darwin's methodology and evidence, Dawson saw no reason to accept his hypothesis over a more benign, teleological explanation. He suggested the more "agreeable" thesis "that each species finding its means of subsistence and happiness constantly extending, exerted itself for their occupancy, and so developed new powers." This explanation accounted for "elevation," insisted Dawson, "as if nature, like a skilful breeder, were giving constantly better food or pasture, instead of imitating the luckless experimenter who strove to reduce the daily food of the horse to a single straw."[68] (Here Dawson would seem to have sided with a Lamarckian explanation, as did many "theistic evolutionists" who objected to the materialism implied by natural selection.[69]) But working naturalists had even more to gain, in Dawson's view, if they remained "content to take species as direct products of a creative power, without troubling ourselves with supposed secondary causes."[70] In essence, for Dawson, the *Origin of Species* provided neither sufficiently accurate reasoning nor sufficiently compelling evidence to persuade him to accept so dismal and dispiriting a view of the natural world.

The more Dawson thought about Darwin the more dispirited he became. He decided to write Lyell again, even though his first missive remained unanswered. He wanted to caution Lyell against those who might seek to identify Darwin's "hypothesis" with Lyell's "doctrine of modern causes," a mistake that Dawson later described as a "mischievous misapprehension."[71] He pointed out that Darwin had admitted that the changes he described might be known only by their effects; they could not be seen in progress. He said that Lyell, on the other hand, could point to elevation and subsidence, volcanic ejections, and river sediments. He stressed that Lyell and other geologists reasoned from strictly natural modern causes, declining to leave the realm of "facts" in order to speculate about the reasons for "the appearance and disappearance of species."

On the other hand, Darwin, said Dawson, had now supplied a "form and coherency" to accommodate the speculations of biologists.[72] Although he had constructed an analogy between artificial variation and variability in the natural world, he had offered no valid connecting link to substantiate his assumptions about the formation of species. In Dawson's opinion, Darwin resembled "those geologists who supposed the tertiary shells were the scallops pilgrims had carried in their hats, and the bones of mastodons the remains of elephants from Roman ampitheatres." Such theories, argued Dawson, explained a great deal only because they required one to *suppose* a great deal. Furthermore, he added, the terms "selection" and "struggle for existence" were but different names for the physical conditions of existence, which by themselves were incapable of producing the richness and diversity of flora and fauna on the globe. In contrast, he concluded, the doctrine of fixity of species provided a firm foundation to paleontology, which neither "the extreme views of Darwin" nor even the idiosyncratic theories of Agassiz admitted.[73]

Apparently the two letters and book review disquieted Lyell to such an extent that he sent Dawson one of his few disapproving letters. Although Lyell, too, would countenance Darwin's doctrine with no higher term than "hypothesis", he felt that Dawson had failed to appreciate "the number of very distinct sets of phenomena" to which it offered a solution (Lyell here anticipated Darwin's own assessment of Dawson's review). Dawson's positing of a "*limited* variability" really explained nothing said Lyell; it amounted, after all, to merely an "arbitrary assumption."[74]

The response to Dawson's evaluation of Darwin was more positive elsewhere, particularly in America. His stance won praise from one of the foremost neo-Lamarckian evolutionists in America, Alpheus Hyatt of the Peabody Academy of Sciences in Salem, Massachusetts. He admired Dawson's "first protest" against Darwin's theory as one that might encourage other Americans of "humbler rank" to come forward. He claimed to speak for other officers of the academy in thanking Dawson for expressing concern at "the exceeding haste of the European world to swallow Darwinism without winking."[75]

Somewhat less impressed by Dawson's writings was Hyatt's fellow American, Asa Gray. Although Gray and Dawson had exchanged opinions by letter and through the published record during the early 1870s, no real understanding or appreciation of each other's position had resulted. Dawson continued to reiterate his view that the hypothesis of evolution ran counter to "true science," especially in its "hypocritical and unfair attempts ... to confuse embryology with geological sequence and to frame classifications looking to evolution and not to actual affinity." He maintained that "the teaching of facts" proved neither "derivation" nor, for that matter, even "the immutability of species." He claimed that Gray's address to the American Association for the Advancement of Science at Dubuque, like Hooker's work on Arctic plants, provided "mere snipes in the dark," representing "only metaphysical speculations aiming to connect isolated facts, the real connections of which are unknown."[76] Happily, said Dawson, this "terrible doctrine" had not been proven true, or as a scientist he would have to believe it, "even though it should rob me of all I value most in this life and that which is to come."[77]

In Dawson's opinion, Gray's endorsement of Darwin's views was dangerous because of their easy extension from the natural world to the social environment. The result of this misapplication, he argued, was "such monstrous creations of fancy" as "the materialistic evolutionism" of Herbert Spencer and Tyndall, as well as "the wretched superstition that man is a descendant of apes" – a belief that "lay[s] the axe at the root of all that is valuable even in the social organization of society." According to Dawson, Huxley and his allies posited that "man is a mere automaton; all his notions of virtue, vice, responsibility, and immortality are mere delusions."[78] He argued that not only did Darwinism sweep away both Christianity and natural religion, but that a populace imbued with "the doctrine of the struggle for existence" would "cease to be human in any ethical sense, and must become brutes or devils or something between the two." Dawson already saw signs of such immoral behaviour appearing among "the lower strata of society in Europe."[79] It was hardly surprising, in Dawson's opinion, that "average humanity revolts against these doctrines."[80]

Gray, for his part, used a review of the *Story of Earth and Man* as a vehicle for disputing Dawson's views about evolution. Dawson maintained his belief in fixity of species with "earnestness, much variety of argument and illustration, and no small ability," said Gray. Yet, as Gray correctly observed, Dawson was convinced that evolutionary doctrines were not only false, but "thoroughly bad and irreligious." In Gray's view, however, Darwin need not be accused of eliminating design from nature. His studies of climbing plants and the fertilization of blossoms by insects had "brought back teleology to natural science, wedded to morphology and already fruitful of discoveries."[81]

But Dawson maintained that, even in the hands of a balanced and reasonable proponent such as Gray, Darwinism was a dangerous doctrine. By the late 1870s – nearly two decades after publication of the *Origin of Species* – Dawson was still outspokenly opposed to the theory of evolution. His longtime London correspondent Bigsby praised him for his "strong, distinct, and trenchant" opposition to the "sad hypothesis" of evolution.[82] In the United States, paleontologists young and old (such as professor Alexander Winchell at the University of Michigan and Charles Doolittle Walcott, future director of the U.S. Geological Survey) applauded Dawson for his continued opposition to materialist explanations.[83] Cornell geology professor Henry Shaler Williams esteemed Dawson as "the one scientific man of reputation in America who has earned the name of not having bowed the knee to the idol of evolution."[84] Dawson continued to be sought by numerous colleges and religious institutions as a lecturer who could respond thoughtfully and intelligently to the onslaughts of the evolutionists. The editor of the *Leisure Hour,* James MacAuley, who arranged for a series of thirteen articles by Dawson, stated that "a page of 'scientific doubts'" from Dawson about Darwinism would be "quoted everywhere."[85]

There seems to be little evidence that Dawson eventually changed his conception of evolution and relaxed his opposition.[86] To him, evolution by natural selection remained a simplistic approach to nature that, by explaining too much, explained nothing. He viewed it as a "philosopher's stone" that "transmute[s] the viler into the more exalted species" or as a panacea developed for "specialists and enthusiasts, who ever tend, like quacks in medicine, to refer all effects to the same cause, and to cure all evils by one specific."[87] Dawson concluded that the history of science was "strewn with the wrecks of such hypotheses, devised in every age by ingenious men to serve as a substitute for actual knowledge, and to spare themselves the labor of arduous investigation; satisfying one generation with a comfortable form of words, only to be cast off by the next."[88] Perhaps he recognized, as one reviewer recently posited, that he could not accommodate his religious beliefs to Darwinian evolution and that the "Christian Darwinism" of Gray and others remained a chimera.[89]

As a result of his trenchant and incessant opposition to evolution,

Dawson began to be viewed as an intellectually isolated figure whose passionate crusades made him an anachronism. Even today, one historian ungenerously paints Dawson as "that most provincial of Presbyterians, ... a solitary figure holding forth from the heights of Montreal until he and his century expired.[90] Although, as one biographer wrote, Dawson did not advance "the trite excuses of the bigot," he did assume the position among evolutionists that Cuvier possessed among naturalists – respected but wrong.[91] Another writer saw him as the last survivor of the pre-Darwinian group of naturalists headed by Agassiz.[92] Even Dawson's wife fretted that he had become too dogmatic and complacent, and stood "*absolutely* alone" among scientists in his defense of revelation.[93] But Dawson, fully convinced of the rightness of his course, proudly wore the mantle of intellectual isolation, as he was to do in other scientific controversies.

10 "A quiet middle course"

Dawson's active and sustained denunciation of Darwinian evolution began to serve as a *modus operandi*. He came to embrace and promulgate – or, alternatively, to reject and denounce – varied scientific theories or hypotheses with marked vehemence, and maintained these positions for decades. His scientific explanations almost acquired the characteristics of religious doctrines: rather than being presented coolly and dispassionately (as he had insisted on as a young geologist), they instead required either a zealous defense or a vitriolic renunciation.

At times, Dawson's work on certain paleontological fragments or geological formations could be used to buttress an anti-evolutionary stance. Certainly his interest in fossil plants of the Devonian stratum, Carboniferous air breathers, and *Eozoön*, dealing as they did with the first appearance of a particular organism in geological time, would seem to have placed him firmly in the arena of evolutionary debate.[1] But usually Dawson paid scant attention to such an application or extension. Of *Eozoön*, for example – heralded at first by the evolutionists as support for their theory – Dawson wrote: "I have avoided all reference to the possible bearing of the discovery on Darwin's views, simply because I do not see that it weighs much one way or the other."[2]

Despite this surprising reluctance to deploy all the weapons in his intellectual arsenal, Dawson apparently loved to plunge into the heat of scientific controversy for its own sake. (O'Brien quips that it served to dispel the chill of Montreal winters.[3]) For this reason, his autobiographical reflections about his labours in geology and allied sciences are astonishing, with their desire for moderate, rather than bold and extreme, views. In the words of Dawson: "I have therefore striven ... to follow a quiet middle course, which, however unattractive to the sensation-loving public, is most likely in the end to be correct."[4] Only in the realm of fossil botany did Dawson seem to anticipate controversy and "some sharp criticism," because opinions are "so conflicting at present."[5]

Besides Devonian plants, two other areas of geological interest preoccupied Dawson over the years, allowing him to indulge his propensity for controversy. These were his desire to challenge those geologists who believed in the efficacy of glaciation, stemming from his study of Post-Pliocene geology; and his insistence on the organic nature of *Eozoön*. The first of these, glacial geology, began to intrigue Dawson in the late 1850s, and was taken up again during the 1860s and early 1870s. Indeed, his decision to study the Post-Pliocene formations around Montreal, Quebec, and the lower St. Lawrence was made on his arrival in Montreal, when he found that other areas of investigation, including Silurian fossils, were already being capably examined by the staff of the Geological Survey.[6] As for the controversy over the nature of *Eozoön*, it became a principal concern of the 1870s, although he continued to address the issue until his death in 1899. Thus Dawson's cultivation of these two areas to some extent follows a chronological progression, although he was to write repeatedly on these topics for decades. J.D. Dana, the seasoned editor of the *American Journal of Science*, always welcomed Dawson's communications on either of these subjects.[7] In both cases, Dawson used Canadian materials – previously unknown and unappreciated by an international community – as the basis of his arguments.

Whether treating glacial action or *Eozoön*, Dawson preached a return to scientific fieldwork rather than armchair reflections; he insisted that fossils needed to be examined *in situ*.[8] The celebration of fieldwork was related to the fact that Dawson saw his opponents as practitioners of faulty scientific methodologies.[9] In particular, he felt they failed to proceed according to the principles of sound induction, which required reasoning based on solid empirical evidence. He, whose scientific apprenticeship had been served by roaming the Nova Scotia countryside, had collected geological and paleontological facts unadulterated by prevailing scientific theories. When being accused of "wild speculation" by an adversary, Dawson responded that he could hardly believe that "the laborious collection of such specimens, the preparation and study of hundreds of slices, and the comparison of them with the forms, recent and fossil, which they may be supposed to resemble" could be stigmatized in this way.[10]

Apart from methodological objections (coloured more often by his religious beliefs than he was wont to admit), Dawson's contentiousness stemmed from his identification of the triumph of evolutionary and related scientific explanations with the waning influence of scientific generalists such as Lyell and Bigsby. By the last quarter of the nineteenth century, the hegemony of the materialists and specialists was emerging – "the reign of the pigmies," as Dawson called it.[11] Writing to fellow American naturalist Asa Gray, for example, he wished that their colleagues across the ocean might "pick up even a *rag* of the mantle of poor Edward Forbes," one of the generalists, in Dawson's view.[12] In his own case, too, Dawson had had a poor reception to his specialization in Devonian plants; this in turn led to further soul-searching about the legitimacy of the fragmentation of natural history. Even more discouraging, his engrossment in this speciality had caused Dawson to neglect other areas (including Post-Pliocene geology and *Eozoön*), where his labours might have been better rewarded.[13]

RAILING AGAINST THE GLACIERS

Ever since 1840, when Louis Agassiz first expounded his glacier theory, geological circles had been in ferment over the leading role assigned by him to land glaciers. Skeptics in Britain and North America (eventually Dawson would be among their number) attributed greater significance to huge icebergs or "floating ice" for producing the geological phenomena in question.[14] It is hardly surprising that Dawson should join the debate on the side of the "drift-ice theory," given his realization, as a lad, of the enormous power of floating ice along the Nova Scotia coast.[15] Throughout the controversy, Dawson echoed his mentor Lyell's fervent uniformitarianism; surely, he argued, the extraordinary power of the ocean that he had witnessed as a youth, for example, might be equally formidable when extended back into geological time.[16]

From the mid-1850s onward, from his new vantage point at McGill, Dawson was to unearth marine fossils in the vicinity of Montreal that seemed to vindicate his adherence to the drift-ice theory. These formed the basis for several long articles on the subject in the *Canadian Naturalist* during the late 1850s.[17] By the mid-1860s, Dawson had renewed his interest in the glacier theory. It formed the core of his presidential address to the Natural History Society of Montreal in 1864, and, in 1865, he took issue with John Tyndall's remarks on glaciers at the British Association meeting in Birmingham.[18] (Dawson even invoked a watery metaphor – "some rather small men [who] are rising to the surface just now" – to describe the physicist Tyndall and his colleagues."[19]) New fossil discoveries by the Geological Survey of

Canada at Rivière-du-Loup and Dawson's own fieldwork in Europe strengthened his attachment to the importance of floating ice.[20]

In 1872, Dawson published his *Notes on the Post-Pliocene Geology of Canada,* in which he responded to the recent work of Columbia University professor John Strong Newberry (for the Geological Survey of Ohio), that provided new support for adherents of a land glacier.[21] Dawson cleverly parried the evidence gathered by such American geologists as Newberry and Agassiz's protégé Addison Emery Verrill, at Yale, with the findings of his son, George Mercer Dawson, in Western Canada, and the work of David Milne Home, in Scotland. He attracted a coterie of followers in the United States, particularly among amateur geologists such as Isaac Newton Vail of Ohio, the Chicago professor and physician Edmund Andrews, and the Cincinnati lawyer Samuel A. Miller.[22] Typical of the group, Miller praised Dawson for "the ease with which you demolish the absurd glacial theories suggested by Agassiz and advocated and enlarged [on] by Newberry and others" in the supplement to the second edition of *Acadian Geology.*[23]

Dawson continued to promulgate his anti-glacier views throughout the 1880s, particularly during his speeches from the presidential podium at the American and British associations for the advancement of science (and also at the Natural History Society of Montreal). In effect, he assumed the mantle as leader of the floating ice theory in North America.[24] The theory required a deluge, which lent it considerable attractiveness to individuals such as Dawson who subscribed to the Biblical account of the flood (it had become unfashionable, however, to make this connection explicitly).[25] But outside the scientific community, individuals such as the governor-general of Canada, the Marquis of Lorne, supported Dawson's view of the connection between the historical deluge and geological phenomena.[26] Even the recently retired professor of geology at Oxford, Joseph Prestwich, found the topic of glaciation and the tradition of the flood to be intriguing.[27]

Once again, Dawson criticized by taking issue with the methodology of his opponents. He believed that the glacialists, like the evolutionists, lacked critical factual evidence in the fossil record to establish their views. He referred to Agassiz's "imaginary glaciers," calling them "physically impossible".[28] He chided American entomologist Samuel Scudder concerning the "glacial poets" who "fabulously reported" unsubstantiated events in the geological past.[29] Elsewhere, he complained of "geologists too much addicted in this matter" who invoked the aid of "improbable causes."[30] He insisted to the American geologist Henry Carvill Lewis, that the "doctrine of the Continental ice sheet" was now "waning," even though most of Lewis's colleagues embraced a theory of successive glaciation at this stage.[31] Capping his nearly forty-year engagement in the controversy over glacial action was his publication of *Canadian Ice Age* (an expansion of his *Notes on the Post-Pliocene*) in 1893.

In the *Canadian Ice Age*, Dawson reiterated his uniformitarian perspective: that he was willing to take account of "the agency both of land ice and sea-borne ice in many forms, along with repeated and complex elevations and depressions of large portions of the continent." This was a view "less sensational than those which invoke vast and portentous exaggerations of individual phenomena" and, therefore, more likely to appeal to "serious thinkers."[32] The conclusion of the work, however, allowed him to consider the implications of his analysis for "larger biological and cosmical questions."[33] The most striking fact, in his view, was that even during the long, climatically and geographically varied Pleistocene period, no evidence had been found to support the hypothesis of the variability of species. Not surprisingly, his analysis of *Eozoön*, as in the case of the Devonian and Carboniferous fossil floras, eventually led him to precisely the same conclusion, and would soon come to replace glacial geology as the focus of his argumentative talents.[34]

EOZOÖN CANADENSE: THE JEWEL IN THE CROWN[35]

Over a period of several decades, Dawson engaged in controversy in a second area, this time involving the puzzling fragment, *Eozoön canadense*, familiarly known as the "dawn animal of Canada." In 1858, a collector for the Geological Survey of Canada found some unusual Pre-Cambrian rocks with alternating "concentric layers of siliceous and calcareous compounds." Survey Director Sir William Logan posited that the rocks – previously thought to be "Azoic", or without life forms – contained organic remains. Most observers remained skeptical when presented with Logan's highly metamorphosed specimens taken from the limestone of several localities in eastern Canada. But when shown some microscopic slices prepared from these specimens, Dawson became convinced of the validity of Logan's views. His identification of the fossil as a foraminifer was confirmed by the leading authority on these large marine protozoans, the Manchester physician William B. Carpenter. Such important support won over two American geologists: James Hall and James Dwight Dana. Even Darwin and Huxley subscribed to this analysis for a time.[36]

Until the mid-1860s, most naturalists remained convinced, thanks to Dawson and Carpenter. The fragments looked like fossils, and virtually nothing was known of pre-Cambrian fossils at the time.[37] But then, as one historian puts it: "This short-lived consensus was abruptly shattered." Two Irish mineralogists, William King and Thomas H. Rowney, published an article in the *Quarterly Journal* of the Geological Society of London on the "So-called 'Eozoonal Rock'." They concluded that *Eozoön* was "solely and purely of crystalline origin."[38] For his part, Dawson found their attack "amusing"; he hoped, however, that Carpenter would not "give [them] an advantage by trying to prove too much."[39] The controversy grew to international proportions in 1866, when the director of the Geological Survey of Bavaria, C.W. Gümbel, announced the discovery of a specimen there similar to the Canadian *Eozoön*, which he christened *Eozoön bavaricum*.[40]

Correspondence indicates that Dawson received important support and information on the *Eozoön* question from his old mentor Lyell, although in this controversy he acted as Logan's "bulldog" (as Huxley was dubbed for his defense of Darwin). Once the dispute became more acrimonious, Dawson's honour, Logan's honour, and above all Canada's honour were all called into question. In Lyell's words, "The Devonian insects and the *Eozoön* throw the old world into the shade"; the authorities in England, said Lyell, have done "little beyond confirming your [Dawson's] views."[41] To Dawson, "The thing is so important that it must be thoroughly worked out, and no space left for doubt either on the part of the mineralogists or zoologists."[42]

Dawson had few doubts concerning the correctness of his views concerning *Eozoön*; by the 1870s, the controversy had moved from national or disciplinary journals into *Nature*, a new nonproprietary journal seeking an international readership. In the columns of *Nature*, Dawson, Carpenter, Rowney, King, and others exchanged salvos about the fragment, with the range of objections to Dawson's identification constantly broadening.[43] Nonetheless, such allies as the mineralogist Thomas Sterry Hunt and the Leipzig geologist K. Hermann Credner stood steadfastly at his side. Whereas Hunt criticized the chemical geology of Rowney and King,[44] Credner attributed the opposition of certain German colleagues to the fact that the Bavarian specimen had failed to reveal its organic structure as clearly and distinctly as the Canadian *Eozoön*. He wanted examples to show to the "doubting paleontologists"; indeed, specimens of *Eozoön* became valuable commodities in the fossil business around this time.[45] One English paleontologist, John Plant, told Dawson that such a "fancy price" had been placed on *Eozoön* by dealers that one needed to be a millionaire to buy any.[46] A London dealer, Robert Damon, initiated an exchange with Dawson, sending him Devonian fossil fishes in return for *Eozoön*.[47] In Philadelphia, Hunt reported that the *Eozoön* specimen attracted much attention at the Canadian exhibit for the 1876 Exposition. Specimens were also displayed at the international exhibition at Paris, two years later.[48]

Even Samuel Haughton, the editor of the *Proceedings of the Royal*

Irish Academy (which published much of the debate), told Dawson that "the weight of opinion" in Ireland favoured the "organic origin of the *Eozoön*."[49] Another British naturalist wondered when Rowney and King will have had "enough."[50] But the debate seemed to drag on interminably, not only in the pages of the Irish Academy *Proceedings* and in *Nature*, but also across the ocean in the *American Journal of Science*.[51] Something of the flavour of the controversy is revealed in an anonymous poem entitled "*Eozoön* Canadense," a "lay of chivalry" penned to commemorate the proceedings in one section of the British Association meeting of 1869 at Exeter. There Robert Harkness, professor of geology at Queen's College, Cork, appeared to throw down the gauntlet by eulogizing Irish geologists in his presidential address.

The Prelude

A marble green one sees
A little like sage cheese,
With metamorphic rocks at rest,
In Connemara, Ireland's west,
And "Serpentine," it will be best
To name it, if you please.

In Canada they've got
Of stones like this a lot,
And scientific writers bold
Wrangle anent these marbles old
Whether the structure they unfold
Be animal or not.

Full hot the wordy fight
Has seemed to ears polite;
But men of science nothing shocks,
And that the oldest, toughest rocks,
Should come in for hardest knocks,
Appears to them quite right.

Lacking a Laureate, he
Whose duty ought to be
To chronicle in tuneful lays
The science-fights of modern days,
A neophyte may win the bays,
In modest poetry.

The Conflict

With harness free from stain and speck,
Canadian Dawson takes the field,
Armed *cap-a-pie* by Beck & Beck,[52]
Crossed hammers blazoned on his shield

And foremost in his train we see,
Unmatched in Protozoan lore,
Great Carpenter, whose chivalry
Has oftentimes been proved before.

And through the lists the cry has flown –
A daring challenge for a fight –
"*Eozoön*, be it known,
Has structure like the nummulite."

And on brave Dawson's gauntlet warm
Is dimly traced in mystic line,
"Outline of film asbestiform,
And chambers all acervuline."

Geologists in silence stand,
Astonished by the challenge bold.
For Dawson's calm and steady hand
Is known and feared by young and old.

But hark! a loud defiant shout
Resounds from Connaught's distant strand,
The "King" of Galway has set out
All mailed in ophite, pen in hand.

And to the field he hies him straight,
With gentle Rowney, knight renowned,
Whose prowess none will underrate,
On Chemistry's broad fighting ground.

Full haughtily they ride abreast,
And as they near the English throng,
Their stern defiance is expressed
In this, the burden of their song:

"Whether from Connemara borne,
Or Canada, 'tis all the same,
For metamorphosed mineral worn,
Is *Eozoön* but a name."

Lo, they have taken up the gage,
And now the chiefs on either side,
Stand armed and eager to engage,
While each his hobby sits astride.

Hot grows the fray, and party cries
Begin to make the welkin ring;
"Strike for our fossil, Dawson's prize" –
"Fight for the doctrine of our King."

Now yield ye, chiefs of Galway bold,
Whilst yet ye are deserving grace,
For scores of Dawson's men untold
Are crowding in the lists apace.

See Burgomaster Gumbel, first
Of Munich's citizens so rich,
And Rupert Jones of Sandyhurst,[53]
And gallant Dr. Anton Fritsch,[54]

Ramsay[55] the great of Jermyn street,
And Sterry Hunt from distant lands,
And many another whom to beat
Ye stand in need of keener brands.

Their harness still is clean and bright
And free from ugly dint and tear,
Whilst yours, with blows from left and right,
Is certainly the worse for wear.

At bugle's sound the knights retreat,
Each to enjoy a breathing space,
But neither side admits defeat,
So we will not prejudice the case,

We merely note that Galway's King
Still chants his old defiant song,

> No fact that Carpenter can bring
> Will prove *to him* he's in the wrong.[56]

In 1875, Dawson brought out a book entitled the *Dawn of Life* (effectively a prose version of the preceding poem), which he saw as describing the "friends and enemies" of *Eozoön*.[57] His decision to dedicate the work to William Logan turned out to be a posthumous memorial, for Logan was to die on the eve of the book's publication. In the book, Dawson claimed that the discovery of *Eozoön*, the "dawn animal of Canada," had ushered in a new era in geology. Dawson and Logan, as well as Carpenter and Hunt, were among the "apostles" of this new era. Whether one believed in a Creator or that "all things have been evolved from collision of dead forces," the discovery of this "last organic foothold" was, argued Dawson, of staggering importance. By poising oneself there, "we may look back into the abyss of the infinite past, and forward to the long and varied progress of life in geological time."[58]

With the *Dawn of Life*, Dawson sought to replicate the popularity of the *Story of Earth and Man*, but the book found few readers among the general public.[59] Perhaps this was because the "popularization" was somewhat half-hearted on Dawson's part. The book had come about, he explained, because it better represented his labour during the past decade than did a series of short, technical papers. But whereas the chapters presented general concerns and even posed rhetorical questions (for example, "What is *Eozoön*?"), each was followed by a long excerpt from a scientific journal to make the work "more complete and useful as a book of reference."[60] By 1876, then, the publisher Hodder & Stoughton was reporting 1000 copies of the book on hand[61]; two years later, attempts were made to persuade Dawson Brothers to take some of the excess.[62] For opponents such as Rowney and King – who had published a long review of the book in the *Annals of Natural History* – Dawson was simply rehashing the same old points of contention.[63] Their dialogue with Dawson seemed to be reaching a stalemate.

The following year, Dawson's position in the *Eozoön* debate changed fundamentally. No longer was he defending or promoting the position of anyone else; instead, he emerged as the sole proponent of the organic nature of *Eozoön*. Carpenter had withdrawn for awhile because of the vituperativeness of Rowney and King's attack, remarking that he would no more attempt to convert the "Galway infallibles" than the pope.[64] Moreover, in February 1875, Dawson's "first patron," Charles Lyell, died.[65] It was a "heavy blow," said Dawson. And exactly four months to the day later, Logan was to die.[66]

Just around this time, the focus of the opposition to Dawson's point of view began to shift: two German zoologists, Otto Hahn and Karl Möbius, displaced Rowney and King as leaders of the opposition to Dawson.[67] Hahn chose to air his findings in the *Annals and Magazine of Natural History,* where, according to Dawson, the editors were looking for a little controversy.[68] Dawson was especially incensed over the inclusion of a second evaluation of his *Dawn of Life,* which he read as "very personal and without a single pretense of any new fact or discovery."[69] Indeed, in this long second review, the obviously self-interested authors, Rowney and King, described Dawson's response to their criticisms as being based on misconceptions, evasions, and "reliance on the already exploded arguments adduced by other writers."[70] Although Dawson had been on good terms with the journal's editor, William Sweetland Dallas, he now wondered whether the publication was about to become a vehicle for "personal abuse and 'startling' reviews of popular works."[71] (Obviously, the inclusion of such reviews would help bolster sales of a usually dry, academic journal.) Dawson welcomed the opportunity to respond to Hahn, who challenged him in his own speciality, microgeology; zoology, he left to Carpenter.[72]

With geological "giants" such as Lyell and Logan no longer alive to help shape the contours of the *Eozoön* debate, the controversy became the preserve of, in Dawson's view, the lesser men: the "pigmies" who possessed large egos despite their small size. Others seemed to agree, the editor of the *American Journal of Science,* J.D. Dana, quipping that Otto Hahn "believes in Hahn"[73]; and William Carpenter reentering the debate as a result of his disgust with the opinions expressed by another Englishman, Henry James Carter, a surgeon in the Royal Navy. Indeed, Carpenter portrayed Carter as "eaten up by egotism," and promised he would receive his just punishment in the *Annals of Natural History.*[74]

Carpenter, like Dawson, understood the *Eozoön* controversy as a methodological debate. In fact, he believed their case to be so compelling that he said he would "abandon microscopy altogether," and present his microscope to King, if he were wrong.[75] Claiming that only those who had decided *a priori* "to exclude all evidence of organic structure" could fail to be won over to their side, Carpenter suggested that the Paleontographical Society appoint a panel of experts (including such naturalists as Huxley, Williamson, and Henry Clifton Sorby) to pass judgment on a "complete monograph on *Eozoön,*" to be coauthored by himself and Dawson. He assured Dawson that this would decide the matter for British and continental paleontologists "in a way that no *Canadian* publication would do."[76]

Dawson and Carpenter responded to this new twist in the debate by peppering their correspondents with *Eozoön* specimens, thinking that good examples would persuade the scientific world of the subject's organic nature.[77] Dawson even recruited his sons to distribute *Eozoön* specimens and William reported that he would deposit a piece at the École des mines in Paris.[78] George Mercer Dawson, chancing on the notice of a small research grant awarded to Rowney and King in the columns of *Nature,* warned his father to expect "something more of a counterblast to *Eozoön.*"[79] Apparently in direct response to this, Dawson applied for a Royal Society of London grant "for the further development and publication of *Eozoön.*" Whether or not he received that support, Dawson intended to spend all his spare money the next summer on "digging out more *Eozoön.*"[80]

Whatever the nature of *Eozoön,* the specimens intrigued naturalists all over the world. Richard Rathbun, then at Yale's Peabody Museum, sent Dawson a Brazilian specimen collected by Orville Derby that resembled *Eozoön.* Such a similarity, in his opinion, demonstrated a geological link between Brazil and North America.[81] Others were more guarded in their thanks, such as William O. Crosby, who, on behalf of the Boston Society of Natural History, responded that he was still "non-committal concerning the animality of the *Eozoön.*"[82] Dawson and Carpenter's strategy backfired, however, when it came to a second German zoologist, Karl Möbius, who moved into the adversarial camp after carefully examining the over ninety *Eozoön* sections given him.[83] O'Brien describes how Dawson tried to discredit his new adversaries, much as he had proceeded with Rowney and King: first, he denied their competence; second, he cited evidence unavailable to them. After shifting the burden of proof to the other side, he confronted them directly.[84]

Forced by Dawson to take an aggressive stance, Möbius demolished Dawson's position in the pages of the *American Journal of Science,* where Dawson had launched an assault against him. Although Möbius persuaded many scientists, neither side could claim a decisive victory.[85] J.D. Dana, editor of the *American Journal of Science,* still remained undecided as the 1880s began, an approach representative of that taken by most of his colleagues throughout the decade.[86] Then, in 1885, Dawson's long-time ally William B. Carpenter died, leaving Dawson as the sole public defender of organic *Eozoön.*

Carpenter's son, Philip H. Carpenter, unwilling to assume the mantle of defender of the organic nature of *Eozoön,* used the excuse that he was too biased. Although Dawson found the rationalization unpersuasive, he did consent to allow Henry A. Nicholson (who had formerly taught in Toronto, but now worked in Aberdeen) to edit the elder Carpenter's *Eozoön* memoir.[87] In the end, however, Carpenter's manuscript was never published. Nicholson, too, seemed to fear taking a public stand advocating the organic nature of the specimens.[88]

By this time, Dawson's own commitment to the *Eozoön* cause was waning. In 1895, he wrote that he had hoped to let the question "repose in peace" and that he had taken pen in hand only to correct some "misapprehensions."[89] Although he continued on several public occasions to espouse his arguments (he published a book about *Eozoön* in 1897, *Relics of Primeval Life*), the focus of his concern shifted elsewhere.[90] The number of his correspondents dealing with matters relating to *Eozoön* – whether asking for specimens or pronouncing on its origins – was reduced to a handful. Perhaps at this late stage in life, Dawson sought refuge in less contentious topics, where he might be remembered for "a quiet middle course."

11 Nova Scotia Revisited

Dawson's scientific agenda was full with such widely ranging topics as Devonian plants, glacial and Post-Pliocene geology, and *Eozoön*. He spent countless hours, as well, on science administration and popularization. In addition, his keen interest in Nova Scotia geology, particularly in the fossils and mineralogical resources of the Carboniferous formation, continued virtually unabated over his years in Montreal.

Contrary to the controversy evoked by much of his other scientific work, there was no disputing the quality and importance of his contributions in this last area.[1] By the mid-1870s, the Dawson family had built a summer home ("Birkenshaw") at Little Métis, Quebec, conveniently located nearly halfway between Montreal and Nova Scotia. There, Dawson could collect specimens from nearby Pleistocene and Paleozoic deposits, or simply return to his "former Acadian hunting grounds."[2] Indeed, Nova Scotia fossils remained a central preoccupation long after Dawson left Pictou, even though duty called him to explore other items: "The truth is that I make the coal and its plants my staple, but ... when other good things offer I leave it for a time. Perhaps this is not the best way for reputation; but it is the best way to be useful here, as otherwise some good things like the Laurentian fossils might remain unworked for a time at least."[3]

COAL REPTILES

Dawson was confident of his skills in the "speciality" of paleobotany; certainly, no one else in North American possessed comparable books or collections on the fossil plants of the Nova Scotia coal.[4] His continuing commitment and interest, however, did not preclude his delight at discovering the occasional Carboniferous animal fossil, although such finds were rare. Dawson's friend and geological colleague, William Logan, is credited with having been the first to discover quadrupeds in the coal formations of Nova Scotia: in 1842, he found reptilian tracks in the rock at Horton Bluff (fifteen miles from Windsor, Nova Scotia). But largely due to the renowned London comparative anatomist Richard Owen's hesitation in confirming Logan's discovery, it was left to Dawson to describe the specimen at length nearly twenty years later. In the interim, Dawson himself discovered fossil tracks in two localities, findings which were communicated to the Geological Society of London in 1843 and 1845.[5]

Dawson was to experience his own problems with Owen. He believed that Owen's *Journal of the Geological Society*[6] publication in 1862 of the discovery of new fossilized reptilian remains was an appropriation of his (Dawson's) own discovery, and that Owen's naming of one of the species after him (Dawson) (*Hylerpeton Dawsoni*) was simply an attempt to placate him. At the heart of the misunderstanding was the fact that somehow the specimens had been misplaced in their boxes and their labels confused. As a result, Owen had "confounded the bones of one genus with the bones of another."[7] This mix-up reflected badly on Dawson, inasmuch as Owen's conclusions contradicted his earlier discoveries. Dawson urged Owen to remedy the problem by issuing a page of errata.[8]

Even though he was dealing with new genera and depending on Dawson's identification of the fragments in order to argue for new species, the ever touchy Owen refused to admit in print that he had erred.[9] He even declined an editorial request to insert errata, saying that they represented only Dawson's "opinion."[10] As a result, Dawson ended up publishing more on the subject under his own name than he had intended. Lyell then entered the fray, insisting that someone like Dawson, who had acted as "an unpaid referee" for the Geological Survey of Canada (particularly by identifying their Carboniferous and Devonian plants) deserved "more than indulgent treatment."[11]

There the matter stood. Several years later T.H. Huxley wrote to Lyell about how "perplexed" he was when having to refer to this matter as a joint production. He complained of Owen's contribution as "superficial and blundering to a degree I had hardly supposed possible." On the other hand, he termed Dawson's work, albeit "worthy and painstaking," as insufficient since he was "no anatomist."[12]

Not content to leave his work on the paleozoology of Nova Scotia in such a fragmentary state, Dawson grouped the fruits of these studies in a series of articles for the *Canadian Naturalist,* that were subsequently published under the title, *Air-breathers of the Coal Period.*[13] This work seems to have been the realization of an earlier wish of Dawson's: to bring "the whole of my Carboniferous geology together in a popular book on the *land* of the Palaeozoic period and its inhabitants."[14] Yet the cost of fulfilling the wish was considerable; as Dawson explained, he was forced to carve out the time from his limited leisure, on the one hand, and from the research time that he preferred to devote to fossil botany, on the other.

Over the next two decades, Dawson continued to examine the fossil tracks of reptiles and amphibians. As one geologist remarks, these studies provided crucial information about vertebrate evolution despite Dawson's own opposition to Darwin's theory. In some cases, Dawson analysed the slabs collected by others, such as the important discoveries of railroad engineer Albert J. Hill. In other instances, with the aid of a grant from the Royal Society of London, Dawson described his own specimens taken from the hollow tree stumps of the coal-measures.[15] As late as 1895, Dawson published a memoir collating his past research on the Paleozoic air-breathing animals.[16]

NEW EDITIONS OF *ACADIAN GEOLOGY*

Dawson's constant updating and tinkering with the contents of *Acadian Geology* indicate his unceasing concern with the paleontological and geological record of Nova Scotia. By 1858, he had already written to James Hall to say that the number of new fossils unearthed in Nova Scotia alone called for a supplementary chapter to the work of 1855. Indeed, he was to produce seventy additional pages in 1860.[17] There, he explained that although he now inhabited "the great Silurian plain of Lower Canada," he still retained a keen interest in the geology of Nova Scotia. For the most part, he added, he was forced to derive his comments on the Post-Pliocene formations, the Carboniferous system, and the Silurian and Devonian rocks from the research of other observers.[18]

Elkannah Billings reviewed this "neat pamphlet" for the *Canadian Naturalist* by providing several long extracts. He concluded by abruptly moving the reader from the level of the more restricted Canadian concerns represented by Dawson's work, to those of the international arena. Ironically, given Dawson's opposition to evolution, Billings referred his readers to the *Origin of Species,* which had been published only the previous year: "Just now when Darwin's theory is attracting so much attention, any organic thing that can be exhumed from such a vastly ancient resting place must possess an extraordinary interest."[19] In Billings's view, then, the excitement generated by Darwin's broad theoretical work spilled over to a narrow empirical study like Dawson's.

The printing of supplementary matter was only a stop-gap measure, however; eight years later (in 1868) Dawson brought out a second edition of *Acadian Geology*. It carried a slightly different subtitle: "The Geological Structure, Organic Remains and Mineral Resources of Nova Scotia, New Brunswick and Prince Edward Island." This time the Macmillan firm replaced Simpkin & Marshall as the London publishers and, reflecting Dawson's presence in Montreal, the firms of Dawson Brothers in Montreal and A. & W. Mackinlay in Halifax replaced the Pictou Dawsons as Canadian publishers.[20]

Acadian Geology had grown to nearly 700 pages (with ten additional chapters) and was illustrated by Dawson. This constituted a major reworking of the original text; subsequent editions would supply only short supplements to this second version. As Dawson later reported, he found that the relevant new material "swell[ed] the work to thrice its original bulk, in spite of all my attempts at condensation, and of the omission of many details accumulated in my notes."[21] Indeed, since the publication of the first edition in 1855, thirty papers on the subject had been published in the *Canadian Naturalist* and the *Journal of the Geological Society of London* alone.[22]

Much of the expanded text of the new edition came from long extracts from other published works, or from manuscripts that had been sent to Dawson for inclusion. Perhaps the fact that Dawson had resided in Montreal for more than a decade by this point necessitated this action; indeed, he depended extensively on the work of others, particularly geologists and mining entrepreneurs in the Maritimes, to supplement the limited research and fieldwork he could conduct during his summer holidays. Certainly, he liberally acknowledged his debt throughout the text, but the presence of so many long extracts and quotations did impart a dry, academic – even wooden – quality to the prose.

Dawson himself was well aware that in trying to serve different audiences with the book, he might not have been entirely successful. He explained in the introduction that certain portions of the books were intended for reference only, and that the reader "when he finds his progress arrested by a dry catalogue, a sectional list, or descriptions of fossils," should "pass on to the next readable portion." Nevertheless, in some sections he did adopt a tone more appropriate for "lovers of the lighter kind of scientific literature," even though these passages on occasion seem forced and inappropriate to the largely academic work. For example, his introduction to one of the chapters in an otherwise tedious discussion of the Carboniferous system is startling and out of place:

> I have already endeavoured to introduce the reader into the jungles and forests of Carboniferous Acadia; but in order that he may fully appreciate the nature of the wondrous vegetation of that ancient time, the producer of all our stores of mineral fuel, it will be necessary that we ... review the several genera of Coal formation plants, and endeavour so to restore them that, in imagination, we may see them growing before us, and fancy ourselves walking beneath their shade.[23]

Obviously the skills of popularization were becoming second nature to Dawson.

The new edition thus updated *Acadian Geology* with regard to recent discoveries in Nova Scotia and New Brunswick, while entering into more general questions of physical geology.[24] Dawson listed eight areas that the new edition explored more fully (or for the first time): the prehistoric human period; glaciation; the Carboniferous flora; the Devonian flora of New Brunswick; the land animals of both periods (Carboniferous and Devonian); Nova Scotia gold deposits; the primordial fauna of Southern New Brunswick; and various actual fossils.[25]

In the second edition of *Acadian Geology,* the contents of the first three chapters remained the same. They introduced the general subject of the volume and dealt with the modern geological formations of "the Acadian Provinces" (the previous year's Confederation of Canada had resulted in this new terminology). Many points were developed more fully in the new edition, such as the term "Acadia." Dawson bemoaned the fact that this beautiful word, which signified a place of plenty or abundance, had been abandoned for "Nova Scotia" and "New Brunswick," the names of the new Maritime provinces.[26]

An entirely new fourth chapter continued the treatment of the modern period by discussing prehistoric man: specifically, the Micmac Indians whose "Stone Age" had occurred just three hundred years before. New sections in the fifth chapter – now entitled "The Post-Pliocene Period" instead of "The Drift Diluvium or Boulder-Formation" – examined glacial phenomena and marine shells. No doubt, the latter topic was included as support for Dawson's hypothesis (dating back to the first edition) concerning the efficacy of floating ice and the Arctic current to account for the deposition of rocks and boulders, as well as the occurrence of striations in rocks. He rejected the glacier hypothesis of Louis Agassiz, A.C. Ramsay, and others, as requiring "a series of suppositions unlikely in themselves and not warranted by the facts."[27] The ninth chapter, although only three pages long, was a new discussion of "The Permian Blank"; that is, of the fact that the European Paleozoic system was absent in Acadia and in the rest of eastern North America.

Dawson enlarged his treatment of the Carboniferous period, replacing the original six chapters with eleven, as well as rewriting the text. He paid more attention to coal fossils, with new sections inserted on the remains of aquatic animals and the Carboniferous fossils of southern New Brunswick.[28] Entirely new chapters explored the marine fossils of the Carboniferous limestones of Nova Scotia, the land animals of the coal period (derived from Dawson's *Air-breathers of the Coal Period* of 1863), and coal flora. Dawson also expanded the discussion of coal-mining, in particular adding extensive details concerning the coal measures of Cape Breton. As he explained, the first edition had summarized the fruits of his work on the government-supported survey of the district of Richmond; by this time, the mines of Cape Breton were far more important and productive.[29]

Dawson enlarged his discussion of the Devonian and Silurian formations from one chapter to four, examining the lower Silurian for the first time. As he explained in the introduction to these chapters, the growth of geological knowledge in Nova Scotia and New Brunswick had been dramatic, as demonstrated by the two meagre chapters in the first edition that had "sufficed for all the rocks older than the Carboniferous"; now, however, "the quantity of matter of these rocks will be more than doubled, and it will be necessary to subdivide them into several series." He also removed himself from the controversy abroad surrounding the nomenclature of the lower rocks, observing that "in America the existence of a great mass of sediment, characterized by a distinct fauna and flora, between the Carboniferous and Upper Silurian, is a fact which cannot be set aside," whatever the disputes concerning Devonian rocks in Devonshire and Ireland.[30] (Later, in the third edition, Dawson said that he had now adopted the "Murchisonian nomenclature ... though under protest.")[31] Nevertheless, he noted that the Devonian strata were irregularly deposited and developed in the Maritimes, with seemingly more interesting fossiliferous rocks in New Brunswick than Nova Scotia. Dawson extracted long passages from the studies of Loring Woart Bailey and George Frederick Matthew for most of the text of one chapter, reserving a full discussion of the Devonian flora of St. John for another. Dawson's Canadian boosterism remained intact: just as many of the Carboniferous fossils of Nova Scotia were unequalled elsewhere, so, too, was the Devonian fossil vegetation of New Brunswick "more perfect, perhaps, than that to be obtained in any other known locality."[32]

In the introduction to the chapter on the Devonian flora – which for the most part listed and described species – Dawson explained the great significance of the formation. The rich plant life of the Carboniferous period appeared to die away during the Devonian; great interest was therefore shown in the few fragmentary remains of plants unearthed in the Upper Silurian. But of the ninety-three species discovered in the area extending from northeastern United States to eastern Canada (placing the area well above Europe in terms of number of Pre-Carboniferous land plants), more than half (fifty) were found in the St. John beds.[33] In this edition, Dawson referred to Murchison's "Silurian system," although he admitted that Acadia could not "claim to be a typical region for any of these [Silurian] series of rocks."[34]

Dawson, did not shirk, however, from coining his own nomenclature to describe what had previously been called the "St. John group." Arguing that the term was too ambiguous to accurately designate what appeared to be the oldest Paleozoic formation yet discovered in North America, he proposed the name "Acadian Group" in its place.[35] The discovery of these and additional "primordial fossils" by Loring Weart Bailey and others permitted Dawson to discern an outcropping of Laurentian rocks in the Acadian region.[36]

Dawson supplied two conclusions to the second edition of *Acadian Geology*. One, a semipopular "Summary of the Geological History of Acadia," recreated the forces that had shaped the geological strata of the region from ancient times to modern day.[37] The second dealt with the extent to which contemporary controversies in geology impinged upon the Acadian formations.[38] Here, Dawson summarized his views on glacial theories, metamorphism, and transmutation of species. Writing nearly a decade after the publication of the *Origin of Species* (by that time, most scientists were persuaded of its validity), Dawson argued that even in the "remarkably complete series of fossils afforded by the Carboniferous of Nova Scotia" he could find no evidence supporting evolution. Furthermore, he said, the distinct relationship between Carboniferous and Devonian flora seemed to militate against it. According to Dawson, the extensive knowledge of these formations made it impossible to fall back on "the imperfection of the record as an argument favouring evolution," although he remained a fervent advocate of Lyell's geological uniformitarianism.[39]

Upon receiving the volume, Lyell sent several long letters back to Dawson during July and August 1868. Flattered by the rededication of the volume to him, he told Dawson that it served as "a grand monument of your labours."[40] His praise became fulsome with further reading of the volume; he concluded that it was "full of original observation and sound theoretical views," which would make it "highly prized" by its readers.[41] He also claimed to be especially impressed with Dawson's sketch of the botany of the Carboniferous plants, calling it a masterly reconstruction of specimens.[42]

Lyell's remarks show that he preferred to deal with the general issues raised by the work, rather than the particular issues of local geology. Nevertheless, he defended the importance of this special locality, noting at the outset that, as Dawson had shown, "it is remarkable ... in how many points Acadia is prominent in geology."[43] Lyell continually sought support for, and parallels with, his own views in comparative stratigraphy, especially because of their recent promulgation (in the tenth edition of the *Principles of Geology* [1867-68] and the sixth edition of the *Elements of Geology* [1865]). He took issue with Dawson only over the latter's interpretation of the characteristics of the glacial period, disputing the hypothesis that icebergs and marine currents were powerful enough to scoop out the basins of the Great Lakes.[44]

Dawson had hoped to use the publication of the second edition of *Acadian Geology* to advance his bid for the Edinburgh principalship.[45] He accordingly printed up circulars containing laudatory excerpts from reviews in leading periodicals as support for his candidacy. Unanimously, reviewers acclaimed the volume as so enlarged and improved as to be virtually a new work. The *New York Evening Post* contrasted this "thick, pretentious octavo" with the "unpretending little volume" that was the original edition. The *Geological Magazine* calculated that it contained five times more material than the original work.[46] In Canada, mining entrepreneur Marshall Bourinot of Cape Breton called the work a "labour of love," which nonetheless contained useful practical information.[47] In New Brunswick, Bailey spoke of reserving it for his best geological students.[48]

More than half the volume – running in this case to 370 pages or eleven chapters – dealt with the "famous Carboniferous formation of Nova Scotia." Yet the reviewer for the *Pall Mall Gazette* hastened to explain that Dawson had applied local observations to questions of general scientific interest. In words so reminiscent of Lyell's letters that one wonders whether he may have been the author, the writer objected only to Dawson's explanation regarding the power of "sea ice" which, he felt, was expressed in excessively dogmatic language and revealed Dawson's weakness in physical geography. Nevertheless, the reviewer observed that *Acadian Geology* was "a work of which the colony may very justly be proud."

Like Lyell, the writer for the *Geological Magazine* commended Dawson for his restoration of paleobotanical fragments, noting the especially tricky puzzles they posed for paleontologists. He explained that in contrast to zoological materials, in which different parts of the organism exist in a definite proportion and relation to one another, the reconstruction of these fragments depended more on imagination than observation. Paleobotanists like Dawson, he continued, had to work with few and imperfect specimens, and without precise, established relationships between the size of fruit, flower, leaf, branch, trunk, and root.[49]

In 1875, a writer for the *Popular Science Monthly* insisted that *Acadian Geology* "still remains the standard work on the geology of the Maritime Provinces."[50] But three years later, in 1878, Dawson brought out a third edition. Actually, it was a one-hundred-page supplement of additions and corrections to the second edition, which could be bought separately (for those who already owned the work) or bound together with the second edition.[51] This time the name of a publisher in the United States was added to the title page: the Van Nostrand firm of New York City.

In the introduction, Dawson explained that, as for the second edition, he had been faced with a "superabundance of matter" in bringing the volume "up to the present state of knowledge." He proudly asserted, though, that his earlier theoretical views required little retraction. In particular, he mentioned that his "position of moderate uniformitarianism, with due allowance for intermittent actions which may well be termed cataclysmic" now characterized the views of a majority of geologists. He also asserted, perhaps with somewhat less conviction, that "the battle of the glacialists" now leaned toward his eclectic view of "marine submergence and ice-drift with local land glaciation." He noted other areas, as well, where geological opinion had swung to his side: "the origin of coal and the land conditions of the Carboniferous age"; "the reality and extent of the Devonian age"; and "the extension downward of the Palaeozoic fauna in America, as far as the Lower Cambrian." Dawson claimed that even the "extension of animal life backward into the Eozoic age" was steadily gaining converts.[52] It is interesting, though, that he avoided mention of his anti-evolutionary stance, an increasingly unpopular and reactionary position among scientists.

He listed his own series of publications dealing with Acadian geology since the date of the second edition, although he admitted that his work had been "limited by distance and by other occupations." He noted his resulting dependence on the assistance of others in the field – including local geologists and collectors, but especially the corps of the Geological Survey of Canada, newly arrived in the Maritime provinces since Confederation in 1867. He explained that he had returned twice to Nova Scotia in the interim, where he was delighted to find the geology of previously inaccessible regions now exposed by new railway lines and mining operations.[53]

Most of the supplement took the reader stratum-by-stratum through the geology of Acadia. Unlike the previous two editions (which Dawson admitted had caused some referencing difficulty), the discussion of mineral resources appeared in a distinct section. Dawson took pains to correct the earlier rendering of the geological map of the Maritimes, while supplying other miscellaneous facts, such as notes on the discovery of Micmac antiquities in Nova Scotia.[54]

The publication of this third edition, with its update of information on the region, almost coincided with the silver anniversary of the first edition and won the praise of Dawson's geological peers. George Frederick Matthew, for example, wrote from New Brunswick that "notwithstanding your absence from these provinces you still hold the honored place of leader in the study of their geological structure."[55]

The only negative note came from Dawson's publishers: the book simply was not selling as they had hoped. Oliver & Boyd complained that the market in Britain was too limited, adding that a Canadian publisher such as Dawson Brothers should be able to find a larger readership. Even the Macmillan firm in London – which could employ marketing strategies beyond the scope of smaller publishers – only managed to sell about a dozen copies a year once initial sales had tapered off. With a surplus of some 500 copies, they advised Dawson that expanding the subject matter to include the rest of North America was unlikely to secure more purchasers in such an inelastic market. They concluded that "all geologists speak of it with utmost respect, but most ... content themselves with [an] occasional reference to it in libraries.[56]

Once Macmillan had placed the remaining copies of the book on the auction block, Oliver & Boyd recounted the publishing history of *Acadian Geology*. The first edition had sold all 1050 copies by 1868, the date of release of the second edition; the second edition had sold 1077 copies prior to 1878, when the supplement was introduced; in the intervening six years, 531 copies had been sold.[57] (These are impressive figures by anyone's standards). Oliver & Boyd concluded that there was sufficient cause for further scientific publications by Dawson. Dawson, incidentally, managed to recycle the woodcuts, using them as illustrations for his handbook of Canadian geology.[58]

In 1891 (nearly forty years after publication of the first edition of *Acadian Geology*), a fourth edition was issued. Perhaps this launch was conceived as a way of using up the third edition volumes, for this latest version simply involved a thirty-seven-page appended supplement. The new title was *The Geology of Nova Scotia, New Brunswick, and Prince Edward Island or Acadian Geology*, with the notion of "Acadia" now clearly in a subordinate position. Yet the seventy-one-year-old Dawson sentimentally opined that this second supplement would update this "labour of love," it being "a contribution to the material interests and scientific reputation of my native country and its sister provinces by the sea."[59]

He began by listing the works of individuals who had contributed to this revision, including official reports by members of the Geological Survey of Canada and publications of independent researchers. Dawson, himself, had published ten papers on the region since 1878, the date of the third edition of *Acadian Geology*. He summarized the supplement as "throw[ing] the new facts which seem most important into the form of a sketch of the general results, and [making] a comparison of these with the geological features of the other similar districts on the East and West of the Atlantic."[60] He compared the contents of the first supplement with that of the second in a table, treating each geological formation in succession; he also supplied long extracts and condensations from the published record (from his own work and that of others) within the text.

Dawson introduced the first section, on the Pleistocene, by saying that although a vast amount of material had appeared on the "so-called Glacial Period," he found little reason to modify the conclusions of the previous supplement. He cited Robert Chalmers' work on the Maritimes, as well as studies of other regions in Canada, to support his view of no "continental ice-sheet, but rather several distinct centres of ice-action." He capitalized on his differences with geologists in the United States by sounding a note of nationalism: these were, he asserted, "the only rational conclusions which can be propounded with reference to the facts observed from the parallel of forty-five degrees to the Arctic Ocean."[61] Here and elsewhere in this short supplement, however, Dawson seemed especially concerned to continually present his views alongside the conclusions of British and American geologists. Paradoxically, although he might be willing to endure intellectual isolation in the international debate over evolution and *Eozoön*, it was an altogether different story when it came to his empirical researches into the geological formations of his native province and country.

Although comparisons were valuable, Dawson contended that Acadia and Canada nonetheless presented unique terrains to the geologists, paleontologists, and mineralogists. He concluded by asserting that the importance of Acadia's geology consisted especially "in that diversity from the deposits of the internal plateau of Central Canada and the interior of the United States." Our "arrangements and nomenclature," in turn, said Dawson, must recognize this diversity, especially inasmuch as Canada is striving to establish its identity *vis-à-vis* the mother country, on the one hand, and its powerful neighbour to the south, on the other. Nothing was to be gained "by conventional arrangements overlooking these differences"; indeed, "separate classifications must exist for the different kinds of areas, in their details."[62] Dawson thus began, in this edition, to use the American term "Ordivician" synonymously with the conventional "Cambro-Silurian," as well as the more controversial terminology "Huronian" and "Laurentian," coined by the staff of the Geological Survey of Canada.[63]

MINING

In the fourth edition of *Acadian Geology,* Dawson reflected that anyone now visiting the coalfields of Nova Scotia and New Brunswick "would go with advantages altogether unknown at the time of my earlier explorations." The mining industry had expanded, he said, with annual output having increased to nearly two million tons of coal; consequently, the region had developed and grown.[64] Originally, Dawson had anticipated that the demands of academic life in Montreal would preclude his further involvement in this expansion of mining and surveying.[65] If anything, however, his departure from his native province had strengthened his ties with a coterie of mining entrepreneurs and intensified his personal involvement in their undertakings.

Richard Brown, director of the General Mining Association (GMA) saw Dawson as someone who would "do anything in your power to promote the advancement of the mining interests of your native country."[66] Another transplanted Nova Scotian perfectly described this attitude: "All true Nova Scotians, whether at home or abroad, are interested in the present and future welfare of Old Acadia through the development of its rich resources in mines, minerals, [or] fisheries."[67] Correspondents in diverse localities therefore sought Dawson's opinion about all aspects of Nova Scotia mining. The problem remained, however, that most Nova Scotian capitalists were unwilling to invest in their province. As the *Pictou Advocate* complained, their "lethargy" was "incredible": "Nova Scotia has to rely on British, American or Canadian capital for the development of her vast mineral resources, while *six per cent* [interest] and Bank Stock engross all the energy and means of our Provincial millionaires."[68]

Local inertia notwithstanding, most regions of Nova Scotia, such as Cape Breton and the area around Pictou, tantalized entrepreneurs with their rich coal deposits. When the GMA's blanket thirty-year monopoly over Nova Scotia expired in 1858 (excluding mines that it was already operating in the most productive regions), new operators moved in.[69] With more than thirty new collieries established during the next few years, coal production increased by more than one hundred percent (from 267,496 to 566,779 tons per year).[70]

But market conditions changed dramatically with the abrogation of the reciprocity treaty between the United States and Canada in 1866. That virtually ruined the coal trade in Pictou, despite the fact that the GMA had sunk new pits to mine a top quality coal.[71] Richard Brown seems to have been the first to ask Dawson about the feasibility of establishing an ironworks there (thereby using the unmarketable coal), as long as high quality, inexpensive ore could be shipped from Ontario (local ore contained only 40 percent iron, or so it was thought at the time).[72]

Although Dawson had left Nova Scotia fifteen years earlier, it was widely believed that he still had "a great deal of influence in the Pictou district," particularly in the event of helping to promote the development of its iron deposits.[73] When the secretary of the Nova Scotia Society for Encouraging Home Manufactures wrote to inquire about the prospects of developing a native iron industry, Dawson's physical distance from Nova Scotia dictated that he refer him to the relevant pages of *Acadian Geology*.[74] Yet Dawson was also chary of acting as an unpaid consultant. He rebuffed the enquiries of George H. Dobson, secretary of the Cape Breton Board of Trade and a member of a House of Commons committee, saying that the answers would require more time than "judging from my past experience, the House of Commons would care to pay for."[75] Now that his scientific and technical expertise had deepened, and the time available for these pursuits had been so drastically curtailed by administrative responsibilities, Dawson was no longer willing to supply free advice.

Whatever the richness of Nova Scotia's natural resources, their exploitation remained a highly speculative activity. So many variables interacted beyond the control of investors and entrepreneurs. As had happened in the case of coal, for example, political expediency – as shown by the reciprocity treaty and its subsequent revocation – could change the face of the market overnight. General economic conditions, the construction and development of new avenues of transportation (such as railway lines), as well as the level of supply and demand flowing from other competitors and consumers throughout the world – all were virtually imponderable.[76] One Nova Scotia lawyer, Robert Haliburton, who had earlier been involved with the New Glasgow iron mines before moving to London, insisted that "it would take all the coal and iron mines in British America to induce me to face again the trouble and worry that these Pictou matters have given me during the past five years."[77]

Nonetheless, in 1872, one mining entrepreneur, Edward A. Prentice, finally managed to pique Dawson's interest in the iron question. He recruited Dawson as a scientific expert, with the promise of endowing a geology chair at McGill or of contributing to one of Dawson's favourite charities. In return, he expected Dawson to survey the iron deposits around Pictou so that "no joint in my armour" would be exposed to a group of British investors likely to take "every precaution that experience, prudence and foresight will dictate."[78] Being assured that the Pictou ores were of sufficiently high quality to make steel, Prentice offered to name Dawson "consulting mineralogist" to the executive committee, and promised "not two but twenty blast furnaces in Pictou before many years."[79]

Prentice's armour was not without chinks, however. Dawson apparently only produced a document with glib assurances concerning the natural and man-made resources of the Pictou area, rather than the hard data that Prentice sought.[80] Prentice complained that he was "somewhat upset" about receiving no assurances as to *quantity*. Prentice had expected Dawson, in his capacity as scientific consultant, to sink at least several trial pits in Pictou, in order to determine the continuity of the iron veins. The economic potential of this undeveloped property had to be more fully proven to these businessmen, he insisted, given the enormous outlay of capital required to build a railway and to erect blast furnaces and rolling mills.[81] When several more pleas for "specific data" about this matter went unanswered, Prentice appealed to Dawson's patriotism. He warned Dawson that "a legitimate development of the wealth of Nova Scotia" was surely preferable to "the formation of a big company got up by Financial Agents who care not a straw beyond the floating of the company, and who would immediately clear out of it, leaving the management in the hands of a lot of ignorant shareholders."[82]

Purely by accident, Dawson seems to have been given privileged information in this matter by his son, George Mercer Dawson. George's employer had offered him a return ticket from London (where he had been studying at the School of Mines) to Canada, if he agreed to help analyze iron ore in the Pictou area, but at the Albion Mines.[83] Writing from Nova Scotia, George described Prentice (his employer's competitor) as being a "troublesome sort of man" who gave everyone the impression of "double dealing" and who had had a falling out with his business partners, thereby risking being sued by the former owner of the ore-bearing lands.[84] The elder Dawson, however, simply urged his son to advise him of the extent of the ore deposits and to "keep a sharp look out for any curious fossils" along the way.[85] Such a naive attitude ill-equipped him to deal with Prentice's business associates: several Scottish investors who looked to drive a hard bargain with the colonists. One of them expressed skepticism at Dawson's claim that the Pictou iron deposits were among the best in the world (rich in "red ore," and well situated in terms of fuel and transport), and insisted that Dawson confess that the asking price of £95,000 was "rather dear." The investor also maintained that besides asking too much, the Canadians were further endangering matters by waiting too long; that is, the iron and coal were already "on the trim."[86]

Matters with Prentice continued to deteriorate, and Dawson became increasingly dependent on his Nova Scotia allies to defend his interests. Long-time friend Howard Primrose reported that he would do all he could to keep Dawson "out of trouble," but that he had "some hard customers to deal with here." At the same time, he urged Dawson to write to Donald Fraser, a Nova Scotia prospector, who had the "highest respect" for him and who might be pressed into service by Prentice. As Primrose concluded, "It would be galling to have others ... reap where we have sown so expensively ... until our crop is gathered."[87]

The solution seemed to be to bring some "Montreal capitalists" into the iron business, a process that Dawson had been trying to stimulate all along.[88] He had already introduced the Molson family – "rich and influential people" – to George, who had been hired to inspect coal deposits for them at Port Hood. (George's mother wanted him in the Molson's good graces because they were "very leading people," but his father urged him only to collect "a good fee.")[89] Dawson also persuaded other Montreal entrepreneurs whose trust he had earned over the past two decades to invest, including the shipping magnate and financier Hugh Allen and Peter Redpath. Probably because of these powerful Montreal interests, the *Gazette* was proclaiming that these were "stirring times in iron and coal, and a lucrative employment for capital."[90]

The Nova Scotians, however, placed only slightly more faith in the Montrealers than the British. Howard Primrose warned Dawson that although the "Montreal Company" held a third-interest in the Pictou ironworks, he did not want to be left in a minority position "with a stranger." Only in native son Dawson did they place "unbounded confidence."[91]

This particular venture concluded by Prentice's threatening Dawson with legal action. A general economic depression a few years later brought all further schemes to a halt.[92] Still, Dawson continued to encourage investment in the Pictou Coal and Iron Company. He simply blamed Prentice's "maladroitness" for the unhappy resolution of this particular issue. Only Peter Redpath seemed to draw a broader moral from the lesson: "It seems impossible to have anything to do with mining enterprises without suffering either in pocket or in reputation, if not in both."[93]

Instead of paying heed to Redpath's warning, Dawson continued to dabble in Nova Scotia mining speculations. The next object of intrigue involved the discovery of copper at Polson's Lake in Pictou County. Immediately, Dawson found himself at the centre of a storm of controversy. Anna Grant, the widow on whose property the copper had been found, asked Dawson to use his influence against the unscrupulous business and government interests poised to take away what was rightfully hers.[94] Dawson's representative in Pictou, W.J. Ross, complained of daily threats of violence against him.[95] Although the dispute was ultimately decided in favour of Mrs Grant, the incident reveals how tempers could flare over even the hope of mineral wealth. Dawson, being physically far removed from the scene, could only rely on his extensive network of friends and colleagues in Nova Scotia.

The discovery of gold provided an even stronger enticement to speculators.[96] It prompted Joseph Howe to ask Dawson about the logistics of establishing a geological survey of the province. With the enactment of Confederation in 1867, however, this task was ultimately subsumed under the responsibilities of the Geological Survey of Canada.[97] The issue was of such interest to Dawson, though, that just days before his death in 1899, he was writing a treatise on gold-bearing rocks.[98]

12 "Mighty trees from small saplings grow"

If Nova Scotians tended to be indifferent toward commercial and financial matters, Montrealers could be faulted for embracing these concerns to the exclusion of higher culture. Such was the conclusion of Samuel Butler when he visited Montreal in the 1870s. His "Psalm of Montreal" commemorates this attitude, using the collections of the Natural History Society of Montreal as an instructive example. In his poem, Butler bemoans the banishment of a classical sculpture of the *Discobolus,* or Discus Thrower, to a room containing "all manner of skins, plants, snakes, [and] insects." The verse reconstructs a conversation between Butler and the society's aged taxidermist, Samuel W. Passmore, who is busy stuffing an owl specimen (itself the symbol of the society):

> Stowed away in a Montreal lumber-room,
> The Discobolus standeth, and turneth his face to the wall;
> Dusty, cobweb-covered, maimed, and set at naught,
> Beauty crieth in an attic, and no man regardeth.
> Oh God! Oh Montreal!
>
> Beautiful by night and day, beautiful in summer and winter,
> Whole or maimed, always and alike beautiful –
> He preacheth gospel of grace to the skins of owls,
> And to one who seasoneth the skins of Canadian owls.
> Oh God! Oh Montreal!

When I saw him, I was wroth, and I said, "O Discobolus!
Beautiful Discobolus, a Prince both among Gods and men,
What doest thou here, how camest thou here, Discobolus,
Preaching gospel in vain to the skins of owls?"
Oh God! Oh Montreal!

And I turned to the man of skins, and said unto him, "Oh! thou man of skins,
Wherefore hast thou done thus, to shame the beauty of the Discobolus?"
But the Lord had hardened the heart of the man of skins,
And he answered, "My brother-in-law is haberdasher to Mr. Spurgeon."
Oh God! Oh Montreal!

"The Discobolus is put here because he is vulgar –
He hath neither vest nor pants with which to cover his limbs;
I, Sir, am a person of most respectable connections –
My brother-in-law is haberdasher to Mr. Spurgeon."
Oh God! Oh Montreal!

Then I said, "O brother-in-law to Mr. Spurgeon's haberdasher!
Who seasonest also the skins of Canadian owls,
Thou callest 'trousers' 'pants,' whereas I call them 'trousers,'
Therefore thou art in hell-fire, and may the Lord pity thee!
Oh God! Oh Montreal!

Preferrest thou the gospel of Montreal to the gospel of Hellas,
The gospel of thy connection with Mr. Spurgeon's haberdashery to the gospel of the Discobolus?"
Yet none the less blasphemed he beauty, saying, "The Discobolus hath no gospel, –
But my brother-in-law is haberdasher to Mr. Spurgeon."
Oh God! Oh Montreal!"[1]

Whether Dawson met Butler during his brief visit to Montreal or knew Passmore is unclear. But when he declined Princeton's call in 1878, he cited (among other reasons) the strong bonds of friendship that had been forged among his circle of friends and colleagues in Montreal during the past twenty-five years. Certainly, he referred to acquaintances made through his church, his family, and McGill University, but he must have been thinking also of those who joined him at the monthly meetings and field excursions of the Natural History Society of Montreal (NHSM). As the motto of the society

("Mighty trees from small saplings grow") proclaimed, great things could be expected from their modest efforts.

On arriving in Montreal in 1855, Dawson confronted an amateur scientific society whose present foundering state belied the promise of its past.[2] Founded in 1827, the NHSM had been the preserve of the city's English-speaking professionals and entrepreneurs: doctors, lawyers, preachers, and merchants, as well as a sprinkling of other vocations. (The society could claim only a handful of French-speaking members.) Instead of merely deriving self-serving pleasure from hobnobbing together, the society members were given multiple and varied rationalizations for their clubbiness. Their catalogues and collections culled from the local environment fulfilled high-minded, if not blessed, utilitarian, political, and religious functions.[3]

The Natural History Society of Montreal's scientific *raison-d'être* had nevertheless fallen into a state of eclipse after 1840, and Dawson is credited with virtually single-handedly revitalizing the organization.[4] The extent of this revitalization has scarcely been fathomed. Before Dawson, the society proceeded in an *ad hoc*, haphazard fashion. With Dawson as a member (and soon as almost its perpetual president), the society's affairs became rationalized and businesslike. Minutes began to display clarity and organization. Where before matters had been left to executive whim, a variety of committees and subcommittees were struck to explore the issues at hand. These included committees to deal with the building, subscriptions, the library, and to lobby the government for a grant. Even the membership was supposed to be divided by discipline: botany, zoology, geology, and antiquities. It thus became the responsibility of members of each discipline to take charge of their respective division of the museum and to furnish papers in that speciality.[5] The society's council and executive committee were enlarged to increase efficiency. A curator, J.F. Whiteaves, was hired to superintend the museum and improve its collections. Earlier, the proceedings of the society had been published in the local newspapers; under Dawson's guidance, the society acquired a journal, where proceedings and reports appeared regularly. The retiring president was given the job of preparing an address concerning the progress of science and the NHSM, to be published in the society's journal.[6]

Less than a year after his arrival in Montreal, Dawson found himself elected president of the Natural History Society of Montreal. A committee (consisting of James Barnston, William Rennie, and W.H.A. Davis) contrasted his "wholesome vitality" to the "most depressing circumstances" of the society's existence, and commended his energetic efforts "to raise the Society from a moribund condition to one of life, vigour and activity." Prior to Dawson's election, it had been difficult to raise a quorum at monthly meetings; even then, most meetings had consisted of general lamenting over the society's debts, whereupon the treasurer would pay them out of his own pocket. Few members honoured their annual subscription. The committee deplored the apathy and "death-like indifference" that prevailed, and especially the "entire absence of any scientific or literary efforts to advance the position of the Society as a scientific institution." Only through the efforts of a few individuals such as Dawson might the society emerge from "its former humiliating condition."[7]

One of Dawson's first acts was to persuade the McGill Board of Governors to provide the society with a piece of land "on terms almost amounting to a free gift" at the northwest corner of University and Cathcart streets, as well as a mortgage to support the erection of a new building. This generous donation permitted the society to move from its "too retired and obscurely situated" location and cramped quarters, where the rooms emitted a "strong, musty and sickening odour." (Small wonder that few travellers and Montreal citizens found reason to visit.) In 1859, the society inaugurated an imposing new building in a fashionable part of town – "a plain but neat and commodious structure, 94 by 45 feet, the style Grecian, with Doric porticoes" – situated in the city's centre.[8] The museum occupied the upper floors.

Typical of Dawson's enthusiasm was a sprightly introductory lecture that he delivered for a popular natural history course, held under the auspices of the society. It was a call to all Montrealers to take up the cause of natural history, since "in truth a large proportion of the new facts added to natural science are collected by local naturalists, whose reputation never becomes very extensive, but who are quoted by larger workers." Said Dawson: whereas "good works of art are rare and costly [like the Discobolus?], good works of nature are scattered broadcast around our daily paths." Dawson proceeded to list "the most promising local fields of inquiry"; the only interruption to his full-blown Baconian rhetoric occurring when he considered the geological domain, which presented the greatest temptation to "vagaries." According to Dawson, the practising geologist would be thankful to receive new facts, provided the amateur did not "oppose the interests of truth by those crude and hasty generalizations, or baseless hypotheses, in which unskilful and hasty observers are too prone to indulge, and which sometimes impose upon the credulity of the public to the serious injury of the science." He cautioned that although no science possessed grander "ultimate truths" and practical results, "none is more dangerous or misleading in the hands of pretenders."[9]

Dawson thereby issued a clarion call to Montreal's citizenry to join in the exploration and enumeration of the treasures surrounding the city, guided by the principles enuciated by the Natural History Society. As the distinguished botanist Asa Gray wrote him around this time, one important function of such a society ought to be to prepare a "full and correct list of the plants of its district." This was not the work of a day or even a year, said Gray, but a combined and continuing effort.[10] He added that the production of a catalogue of the local environment might bring together gentlemen and scientists, old and young, amateurs and professionals.

Dawson served as president whenever no one else could be persuaded to undertake the job: between 1856 and 1890 (when he was elected honorary president for life), he occupied the chair twenty times.[11] On one occasion, when he did not wish to serve again as president, the society prevailed upon him to stay.[12] On another, they suspended the rules so as to acclaim him to the presidency.[13] Even after pleading that he needed to husband his energies as he approached age seventy, he was still unsuccessful in vacating the position.[14] When not president, he often served as vice-president or chairman of the council; he was even willing to fill a chair temporarily for an absent officer.

THE AMERICAN ASSOCIATION FOR THE ADVANCEMENT OF SCIENCE COMES TO MONTREAL IN 1857

Interested in both reinvigorating the Natural History Society of Montreal and attracting international attention to the city's scientific resources, Dawson (then president) hatched a brilliant scheme. He would invite the American Association for the Advancement of Science (AAAS) to hold a meeting in Montreal. Vigorous lobbying by Dawson, Logan, Smallwood, and other Montrealers at the Albany meeting of the AAAS in 1856 led to the acceptance of the invitation, disappointing those from Baltimore and Springfield, Massachusetts.[15]

In Montreal, planning for the impending meeting allowed the society to assume a community presence as never before. Prominent residents formed a large local committee to handle organizational details. Free passage for some participants was secured from railway and steamship lines. Newspaper coverage of meetings and other functions, to be extended and expanded in later years, was obtained. Even the federal government agreed to grant £500 toward the society's new building. Originally, Dawson and others had hoped that the new quarters would be erected before the meeting convened, but in the end only the foundation stone could be laid, albeit amid much congratulation.[16]

On 12 August 1857, the first meeting ever held by the association outside the United States opened under the presidency of Alexis Caswell, professor of mathematics and natural philosophy at Brown University. As he remarked in his inaugural address, "We have left the American eagle, but we assure the gentlemen of Canada that we feel in no danger of being harmed by the British lion."[17] Clearly, the rank and file shared Caswell's sentiment, for, with the exception of Albany the year before, Montreal was to register more members than was the case at any previous meeting.[18] But attendance by leaders in the American scientific community was impressive by any standards. Joseph Henry (of the Smithsonian Institution), Alexander Dallas Bache (of Philadelphia), and the physicist Joseph Le Conte all read addresses to the Physical Science Section. Benjamin Silliman (of Yale) and Benjamin Pierce (of Harvard) lectured to the section of natural history and geology. Leading American geologists also participated in that section, including Edward Hitchcock (of Amherst College), Arnold Henry Guyot (of Princeton), James Hall (of New York), and Ebenezer Emmons (then state geologist for North Carolina). James Dwight Dana, professor of natural history at Yale, addressed the section on ethnography and statistics, as did Silliman in a second paper. Canadian scientists, too, did their best to animate the proceedings with Dawson, Logan, Hunt, as well as Toronto's Daniel Wilson and Edward Chapman, each addressing the assembly on several occasions. A number of British scientists attended as well. A.C. Ramsay represented the Geological Survey; Berthold Seaman, the Linnean Society; and the Arctic explorer John Rae delivered a special closing day lecture on the search for John Franklin.[19] In addition, Edinburgh professor Philip Kelland participated in the meeting.[20]

A number of less serious diversions were arranged to complement the formal sessions. James Hall gave an address, a popular lecture, and NHSM members were given the opportunity to mingle at a special exhibit of prized Indian artifacts. About 800 Montreal citizens attended a conversazione at City Hall on Thursday night. Friday evening entertainment took place on the McGill Campus, in Burnside Hall, followed by an excursion Saturday afternoon to nearby St. Helen's Island. The closing evening permitted a fuller "display of local oratory and fashion," with yet another excursion organized to the rapids of the St. Lawrence and surrounding sights. As the reviewer for the *Canadian Naturalist* concluded, "Tastes differ, and these recreations of Science were as different as tastes; but each was eminently successful, in its own way."[21] Only the Philadelphia mining professor, Peter Lesley, dissented, claiming that the meeting was "badly managed," Hall's address was unfinished, and the governor-general had "left town in a huff."[22]

For the members of the Natural History Society, however, there was no disputing the success and significance of the meeting. To them, it was quite simply "the most noteworthy event" that had ever happened in their history. They believed that the meeting's success would stimulate interest in scientific pursuits, both in Montreal and throughout the province, and possibly even lead to the formation of a Canadian Scientific Association. But no matter what happened, the members had been cheered by the event, as their numbers substantially increased.[23]

DAWSON AND THE *CANADIAN NATURALIST*

Under Dawson's first presidency, the NHSM took over publication of *The Canadian Naturalist and Geologist*. Although two volumes had already been edited by Elkanah Billings, he was happy to be relieved of this responsibility for it took time away from his survey work.[24] Henceforth, the journal would be directed by an editorial committee of the Society (which always included Dawson), and the publisher, John Lovell, replaced by the Dawson Brothers. To Dawson, the *Canadian Naturalist* was the means of transforming the NHSM from a gentleman's club into a serious amateur scientific organization. Not only did it provide an avenue for the publication of the members' scientific work, it meant that the society could exchange its journal of record with institutions elsewhere in Canada, the United States, and Europe.[25] The twin benefits were the expansion of the holdings of the society's library and the international promulgation of the work of its members.

The Dawson Brothers did not simply rubber-stamp the decisions of the editorial board: they kept a tight rein on all aspects of the journal's production. From the beginning, they carefully controlled expenditures, only agreeing to supply woodcut illustrations and a limited number of exchange copies out of the substantial press run of 850 copies. But the *Naturalist* still operated in the red.[26] Various strategies were devised, with the aim of increasing journal sales (and ultimately revenue). A new class of NHSM member was created in the mid-sixties; called "non-resident ordinary members," their entire dues were applied to a *Naturalist* subscription. A few years later, the journal claimed to have achieved a "new and more popular basis," with more variety in its articles; quarterly rather than bimonthly issues also meant substantial savings.[27] In addition, the society's council proposed that the lion's share of its grant from the provincial government go toward the "maintenance and increased efficiency" of the journal.[28]

The Dawson Brothers, still unsatisfied that the Natural History Society was doing everything it could to promote sales, demanded that the NHSM guarantee the purchase of more copies and accede to the appointment of an editor. Otherwise, they said, they would cease to publish the *Naturalist* and forbid any other publisher to use its title.[29] Even under this new arrangement, with Bernard Harrington acting as editor, the incessant bickering between the Dawson Brothers and the society caused the journal to "stick in the mud."[30] The society was ultimately forced to reassume responsibility for the journal, which occurred in 1888 with the publication of the second volume of the *Canadian Record of Science,* successor to the *Naturalist.*

Despite these difficulties, Dawson could boast that here was a Montreal product, for which "the plates are executed and the work printed and bound in this city."[31] The *Canadian Naturalist* – a symbol of the importance of the Natural History Society and its host town – gave Dawson collateral for exchange with leading scientific societies and institutions throughout the world. By sending copies of the journal to organizations as diverse as the Boston Society of Natural History, the Royal Society of London, and the California Academy of Sciences, Dawson obtained their transactions in return; these were then deposited on the shelves of the Natural History Society's library. He also exchanged the periodical for other desirable journals, such as the *Geological Magazine,* thereby augmenting the society's collection. A subscription to the *Naturalist* formed part of one of the hard bargains driven by the Rochester, New York, natural history dealer Henry Ward; it netted natural history specimens for the society's museum in return.[32] Another American dealer, Victor Lyon, sent fossils in exchange for a complete run of the *Naturalist.*[33] By 1870, the journal proudly announced that it was copied *in extenso* in leading scientific periodicals, such as *Scientific Opinion,* a London publication.[34]

Dawson also used the *Canadian Naturalist* as a means of publishing his addresses to the society, communicating miscellaneous information and excerpts from other scientific journals, and disseminating his research and fieldwork. As he had announced from the beginning, geology was to occupy a central place on the society's agenda and dominate the content of the *Naturalist,* in part due to the active role played by officers of the Geological Survey of Canada.[35] But even during the late 1880s, the fruits of the labours of the "knights of the hammer" still eclipsed all other contributions to the *Canadian Record of Science,* successor to the *Naturalist.*[36] (Shortly thereafter, one of the editors remarked that a higher proportion of papers on biological subjects might make the journal more interesting to more readers.)[37]

Rhetorical defences of the importance of geology punctuated Dawson's annual presidential addresses. To members of the society, he promulgated the view that geology "in our day sits justly enthroned as queen of all the natural history sciences."[38] Years later, he still claimed (with customary bias) that geology presented "the largest and most attractive field open to students of nature in Canada."[39] The pervasiveness of this outlook is shown by the scope of the presidential address of amateur astronomer and meteorologist Charles Smallwood, who defended his attention to other departments of physical science because past addresses had taken no account of sciences other than geology and botany.[40]

Dawson's knowledge of the mechanics of science publishing (particularly of the restrictions that leading journals placed on their authors) enabled him to entice contributors to supply top quality manuscripts. Dawson offered the American fossil entomologist Samuel Scudder, for example, the chance to publish a paper in its entirety that the Geological Society of London had agreed to issue only in abstract form.[41] Dawson's son, George Mercer Dawson, published the same article in the *Naturalist* and the *American Journal of Science* – "a method of making a little work go a long way."[42] In particular, Dawson *père* actively recruited the assistance of younger Canadian geologists who were exploring the boundaries of the new nation. Accordingly, Bailey described the Grand Manan islands, George Frederick Matthew discussed his fieldwork in New Brunswick, Francis Bain treated the land shells and molluscs of Prince Edward Island, and Joseph W.W. Spencer explored the geology of the western end of Lake Ontario. Henry Alleyne Nicholson, who supported Dawson's efforts during his short tenure at the University of Toronto, believed that if the *Naturalist* ever ceased publication, it would be "a real loss" to Canada.[43]

Dawson himself unabashedly used the pages of the *Naturalist* (and the podium of the society) to flog his *bêtes noires*: evolutionists, glacial geologists, and those who disputed his contentions about the organic nature of *Eozoön*. As he explained on one occasion, the sphere of the society "as a modest collector and preserver of local facts in Natural History" did not preclude its consideration of "more difficult and abstruse questions which agitate Naturalists elsewhere."[44] For example, he interpreted J.F. Whiteave's dredging of Post-Pliocene shells from the St. Lawrence (expeditions sponsored by the society with the support of the federal government) as supporting the view that species do not change.[45]

THE 1860s: DECADE OF SUCCESS

Dawson's promotion of the *Canadian Naturalist* in particular and the Natural History Society of Montreal in general was both multifaceted and unceasing. In 1859, when the society inaugurated a new suite of meeting rooms, he began to seek out donors to present unusual collections to the museum, or to loan objects from their private cabinets for special exhibitions.[46] He berated speakers who threatened to cancel, saying "Please do not fail us."[47] He encouraged other amateur organizations to collaborate with the society, such as the Numismatic and Antiquarian Society of Montreal, and the "dormant" but "by no means dead or defunct" Microscopic Society of Montreal.[48]

Dawson, who was beginning to perfect the act of going hat in hand to legislative bodies, may well have been the motive force behind the society's quest for a government grant. As early as 1858 (during his first term as president), he wrote in the annual report that there was no other scientific institution in the country "so comprehensive in its aims as ours is, possessing a larger collection of scientific objects than our museum contains, or publishing transactions on natural history of greater scientific value than are to be found in our Journal." His regular complaint became that the Montreal society was unfairly disadvantaged – being treated like a "country Society without a local habitation or a name" – while organizations like the Canadian Institute in Toronto benefited from government largesse.[49] Finally, in 1860, the provincial government awarded a grant-in-aid of $1000 a year.[50]

By the end of Dawson's first decade in Montreal, the society's situation had improved markedly. The series of Sommerville lectures was attracting more auditors than ever before. A vigorous membership campaign had brought in nearly one hundred new members in 1862 alone, increasing the rolls by nearly one-third.[51] And the society would inaugurate the enormously popular annual conversazione the following year.

At these gala affairs, spectators were regaled with displays of microscopes and other scientific instruments, as well as a series of "dissolving views," while the strains of the band of the 63rd regiment of the Royal Artillery were heard from the museum gallery. Tables presented specimens of *Eozoön,* a prehistoric human skull, and bones from a mammoth. Among the apparatus on display was a swinging pendulum, as well as an electric and fire-alarm telegraph.[52] Only the exhibition of chemical experiments was cancelled because of concern that "the gases emitted in the performance of the experiments might not tend to improve the ventilation of the room."[53] Even amid the general economic depression in 1865, more than 400 people still attended the annual conversazione.[54] Soon the details of these affairs were more finely polished: a ladies' committee decorated the rooms; flowers from the private conservatory of a member were exhibited; and David McCord designed permanent ornaments "to recall to mind the names of the leaders in different departments of science, emblazoned with mottoes and emblems."[55] Music increasingly played a role, in the form of the Germania glee club or a church choir, and "Plateau's soap bubbles" were released, charged with gas. Live zoological specimens were introduced, as well as an aquarium with living sea anemones.[56] The eighth conversazione welcomed Prince Albert to its yearly gathering; on that occasion, geological maps and sections lent by the Geological Survey of Canada were displayed.[57]

By the late 1880s, although the frequency of the conversaziones had diminished to perhaps one every five years, they continued to be viewed as "the greatest means of popularizing science in Montreal."[58] Dawson's rationalizing spirit extended to these affairs as well: by this time six subcommittees, including music, decoration, and reception, were struck to deal with the myriad matters engendered by the events' heightened success. Newspaper accounts of the day reported that the NHSM building appeared as "a blaze of light" thanks to its electric and gas illumination, the museum curator having flourished his "magician's wand" to transform the entrance hall by means of "beautiful heads and antlers of moose and buffalo." Although the numbering system employed in the cloakroom was termed especially impressive, the reporter also opined that the refreshments demonstrated that "the savants of Montreal have no contempt for gastronomic science." All in all, the NHSM seemed to have entered a new era of prosperity.[59]

In an even more dramatic display of popular involvement, the society voted in 1868 to open its museum to the public on Saturday afternoons. Initial vandalism persuaded the mayor to supply two policemen for these occasions; even so, no more than 130 visitors attended on a weekend (sometimes as few as thirty).[60] Clearly, the inanimate collections of the museum failed to fire the popular imagination in quite the same way as did the circus-like conversaziones.

More successful were the annual field excursions to outlying areas such as Ste. Anne-de-Bellevue, Ste. Helen's Island, and Mont Ste. Hilaire, during which prizes were awarded to the day's most zealous collectors. A special train (supplied by Van Horne, whose daughter was among the excursionists) took the group to Joseph Papineau's magnificent estate and personal museum at Montebello, where they were joined by a contingent from the Ottawa Field Naturalists' Club.[61] On another occasion, the town of St. Eustache greeted participants with flags, bunting, and banners down its main street proclaiming "Honour to Science" and "Be they welcome."[62] At a fall field trip to Charles Gibb's extensive apple orchards in Abbotsford, Dawson managed to lecture to the geological party from atop Yamaska Mountain, despite a snowstorm during the group's ascent.[63]

Dawson's presidential address to the second conversazione reveals some of the reasons why Montrealers found the activities sponsored by the NHSM so attractive. The society aimed to promote social harmony; as Dawson was wont to say, "A true naturalist is never an ill-natured man." It also cultivated the study of "things that make for peace" and "for the common benefit of all," whatever the prevailing "perturbed social and political elements."[64] Furthermore, the society's umbrella protected against the ill-winds of modern, industrial civilization. As Dawson put it, the members ensure "a testimony in behalf of nature," during "these artificial days." In an interesting metaphor (given the Quebec context and the nearly complete disregard of the French element therein), the society stood up for "the lily of the field" against "all the glory of modern art." Nor were the utilitarian functions newly embraced by the society to be ignored: the society would go to the rescue of anyone "puzzled" by some "unaccountable phenomenon in air or earth;" if "any impertinent insect or fungus ravages your farm, garden, or orchard;" or to those threatened by some "perversion of mining enterprise."[65] Perhaps in order to justify its recently obtained government grant, the society began to take an active role in practical issues around this time: it lobbied for the passage of legislation to protect insectivorous birds, struck a committee to examine the reasons for the decline of apple orchards on the island of Montreal, and started to explore the question of using Canadian fibres in paper and fabric manufacture.[66]

The NHSM also instituted the annual award of a bronze or silver medal to a distinguished scientist (the first silver medal went to Toronto's Daniel Wilson). Ever-expanding collections in the museum necessitated the hiring of J.F. Whiteaves as scientific curator, not to mention the training of the society's janitor, Mr. Hunter, to collect and prepare specimens. Eventually, Hunter was called the "cabinet keeper" and was even sent on collecting expeditions, "an agreeable change from his ordinary employments."[67] Soon, he was being referred to as "our skilful bird-stuffer." In 1870, however, he resigned because of personal reasons and ill health.[68]

FROM BOOM TO BUST

The new face of the society and its range of activities were not achieved without a price. The new building cost more than $10,000, while the old one sold for a mere $2000; the result of this transaction was a legacy of debt that would encumber the society for years. Even after the glorious decade of the sixties, the society remained in debt for more than $2000, the books not showing a positive balance (of $16.80) until 1889.[69] Moreover, the society always incurred a deficit from its popular field excursions and conversaziones. The publication of the *Canadian Naturalist* was a financial drain as well.

By the late 1860s, with membership stagnating and thereafter declining (perhaps as a result of a general commercial depression), the society launched a diligent search for new sources of revenue. The subscription rate was raised to five dollars a year (from four), and "ladies" were sought as associate members for two dollars a year. As well, donors were asked for contributions to a special fund; first, for liquidating the accumulated debt of $2400 and, later, for improving the library holdings and museum collections.[70] Alternative suggestions spoke of affiliating with other Montreal societies, putting all libraries in the city under one roof, and adding a gallery of fine arts.[71] These measures proved somewhat ineffectual, however, given that for the next several years the council recommended "an active canvass for new members."[72] By the mid-1870s, the council decided to transfer custody of both the museum and library to the Fraser Institute, which relieved the society of general maintenance and allowed them to divert their limited funds to improving the content of both rooms. The lecture room was rented out to various organizations, including the horticultural, numismatic, and philharmonic societies.[73] Even the annual conversazione saw a decline in attendance.

Beginning in the 1870s, the Natural History Society of Montreal seemed to fall upon relatively hard times, entering, in Dawson's terms, a "slow and languid condition of our progress."[74] Long-time enthusiasts such as Dawson (as well as Geological Survey of Canada director A.R.C. Selwyn) had little time and energy to expend on society matters.[75] This development is reflected in the increasingly perfunctory proceedings of the meetings (printed in the *Naturalist* and recorded in the minute books) that registered only a few donations to the museum and the reading of an occasional paper. Presidential addresses likewise became brief. Although Dawson used these occasions to address his current passions, such as *Eozoön*, evolution, and nomenclature, such discussions were scarcely comprehensible to most members. Obituaries of many leading members began to appear with increasing frequency in the pages of the *Naturalist,* including those of William Logan, Elkanah Billings, Philip Pearsall Carpenter, and George Barnston. Meetings were plagued by poor attendance (in the late eighties, often only about a dozen members attended) and few members came forward with new data or collections.[76] In 1880, the editors of the *Naturalist* commented upon the "scantiness" of the material received, and urged members to execute more "scientific work" for record in the journal.[77]

Even the society's curator of thirteen years, J.F. Whiteaves, tendered his resignation in 1876 in order to assume full-time responsibilities as paleontologist to the Geological Survey of Canada.[78] His dredging expeditions in the Gulf of St. Lawrence thus came to a halt, as did the subsequent enrichment of the society's collections – all of which had come about in the first place only as a result of Dawson's vigorous lobbying of the government.[79] As well, the provincial government reduced the society's usual grant by 25 percent, started to delay its payment, and eventually withdrew its support altogether.[80] By the late 1870s, the Geological Survey – itself the child of the Natural History Society of Montreal – threatened to move its staff and museum to Ottawa.[81] Although the NHSM registered its outrage at this possibility, the threat nonetheless served to demoralize the scientific community in Montreal.[82] The loss of the museum – "a kind of sacred inheritance to Canada" – appeared as "an act of the grossest vandalism," in the words of Dawson, once again president of the NHSM on the eve of its relocation.[83] It soon became apparent that the absence of such survey staff members as Robert Wheelock Ells, Joseph Frederick Whiteaves, Arthur Humphreys Foord, and Thomas Chesmer Weston would prove to be an even more irreplaceable loss.[84]

RETURN OF THE AMERICAN ASSOCIATION FOR THE ADVANCEMENT OF SCIENCE

Dawson consoled himself for the society's recent lacklustre performance by reflecting that "scientific societies in a country like this are of slow growth." Nonetheless, he believed that "after an existence of half a century, and after having held up the torch of science for that long [a] time in a community, this Society should have acquired greater strength."[85] Among the alternatives that he must have pondered with a view to reinvigorating the society during "its second half century"[86] was the invitation of 1857 to invite the American Association for the Advancement of Science to meet in Montreal. Surely a return engagement might regenerate the flagging enthusiasm for the society's pursuits, among both the members and the community at large. It could also be used to celebrate the golden anniversary of the incorporation of the society, and the silver anniversary of the first Montreal meeting.[87] At the Boston meeting of 1880, the Montreal society duly issued an invitation for the meeting of 1882.[88] By the rules of the AAAS, no invitation could be accepted more than a year in advance; nevertheless, the widespread approval in Boston was interpreted as "virtually pledging" the society to subsequent acceptance at the Cincinnati meeting in 1881.[89]

Indeed, the mere announcement of the AAAS's return to Montreal gave an enormous boost to the Natural History Society's morale – a reinvigoration that only slowly dissipated during the 1880s. The

Canadian Naturalist proclaimed that the "Society is feeling the influence of the good times upon which our country is now entering."[90] The museum, newly opened to the public before the Sommerville lectures, attracted thousands of visitors on those lecture evenings.[91] And city schools and colleges were invited to visit the museum on Saturdays. As a result, the number of annual visitors increased: from 451 (1888), to 1192 (1889), to 2094 (1890).[92] The subsequent extensive renovation, repair, and rearrangement of the museum seemed to bear out the earlier contention that the society might indeed benefit from the removal of the Geological Survey museum, given the reduced competition for the public's attention.[93]

The Natural History Society of Montreal approached the 1890s with renewed confidence. This assurance reflected not only popular support but also the financial well-being that accompanied the abolition of its debt, increased rents from its rooms, and reinstatement of its provincial grant.[94] Still, Dawson continued his vigorous promotion of the society's increasingly broad sphere of activities. Only in 1893, when ill-health associated with a stroke curbed his activities, did Dawson cease to attend society meetings regularly. The then-president, reflecting on Dawson's unusual absence, mused that even when Dawson was so preoccupied with other matters, he had never missed monthly meetings, poorly attended or not.[95] Even as late as 1897, Dawson read a paper before the group. A tally of his contributions to the *Canadian Naturalist* and *Canadian Record of Science* exceeds one hundred and seventy articles. Not without reason was the NHSM to memorialize him as "the mainstay of the Society for upwards of forty years."[96]

13 Putting Montreal on the Scientific Map[1]

Even before the Natural History Society of Montreal's invitation was tendered at the Boston meeting of the AAAS in 1880, permanent secretary Frederick Ward Putnam was excitedly writing to Dawson. Sure that the invitation would be accepted – so many AAAS meetings had recently been convened in the southern and western United States that it was now time for a series of eastern venues – he recalled with pleasure the last Montreal meeting, where he had been the youngest member in attendance.[2] Originally Dawson had thought to extend the invitation for 1881; a postponement to 1882, however, would permit the new Peter Redpath Museum to be completed and the question of the removal of the GSC museum to be settled.[3] Although an official acceptance was only sent to Dawson from the Cincinnati meeting of 1881, where he was elected president, he had, in fact, begun preparations in Montreal months before.[4]

THE AAAS MEETING OF 1882

By the following December, Dawson began to organize the local committees.[5] Chemist Sterry Hunt headed the executive committee, while six special committees – each composed of twenty to thirty men – took charge of everything from printing and publication to conveyance and entertainment. Members were recruited from the city's political and financial elite, including Major S.C. Stevenson, banker Francis W. Thomas, businessman Henry H. Lyman, and former mayor W.H. Hingston. (These were the same prominent citizens who had earlier been asked to contribute to a guarantee fund to cover the expenses of the meeting.)[6] Although Dawson earned Peter Redpath's blessing for tapping this resource, he did wonder whether they would actually set to work: "The meeting is of such importance that I hope the zeal now existing will not evaporate when work is to be done. Will not many of them be away at the coast or elsewhere when the meeting takes place?"[7]

It was Dawson who ensured that local committees performed their stipulated functions; in one case, he sternly admonished the honorary secretary, F.W. Hicks, for neglecting his duties. As for the executive committee, it alone managed to raise $4300 to defray the living and travelling expenses of AAAS members.[8] Although Dawson had delegated responsibility to these eager Montreal residents, he took charge of countless organizational details on his own. As his close friend Daniel Wilson put it, "The outsiders who see everything going along smoothly, have little idea often how much care it has cost to secure this."[9] Dawson's wife preached constantly to her husband about "the folly of wearing yourself out with what you yourself will see to have been trifles"; she advised that "some things evidently unfinished would better show how much had been done."[10] In her view, "Finishing touches may go on forever," and "may be bought at too dear a price." After all, Dawson was, at this point, sixty-two years old.[11] Longtime McGill colleague Alexander Johnson worried about the "continued strain" on him, even given his remarkable "energy" and "strength of constitution."[12]

Dawson's family, of course, rushed to his assistance. His wife, Margaret, helped with the protocol of invitations, while sons George and William (the latter would have "a good opportunity to advance his career," in his brother's opinion) promised to return to Montreal from other parts of Canada for the meeting.[13] Rankine sympathized with his father for his "onerous position ... in the double capacity of president and host,"[14] while son-in-law Bernard Harrington reported that he was receiving ninety-nine "acceptances" for every thirty-three "regrets" each day.[15] Ever concerned with extending a warm welcome, Dawson invited George Brush (the retiring president of the AAAS) and Asa Gray as house guests; he also offered to entertain his longtime correspondent, the fossil entomologist Samuel Scudder. He wrote to Alexander Agassiz at the Museum of Comparative Zoology, saying that John Henry Molson – who had "one of the most charming residences in Montreal" – wanted to be his host.[16] Up until the eve of the meeting, Dawson was bothered by correspondents seeking lodging, special arrangements for their talks, and an opportunity to exchange specimens or scientific arguments with him. A week before his departure for Montreal, Putnam reported that he, too, was "driven to death with Association matters," but that he had fifty-four papers lined up for delivery at the meeting.[17]

Montreal did not, in the end, prove to be "broiling hot," as is so often the case at the end of August; that year the intense heat of July gave way to cooler temperatures.[18] On the evening of 24 August, the Peter Redpath Museum formally opened its galleries with a reception for 2000. Newspaper accounts described the museum as thronged with guests and decorated with flowers that "took all the grimness out of the monstrous antediluvian skulls and skeletons." One reporter claimed that the "principal scientific lights of the nineteenth century" were all on hand, accompanied by wives and daughters, as well as the "elite of Canadian Society." Certainly visiting members of the AAAS dominated the gathering, but McGill University staff, benefactors, and graduates also numbered among the invitees.[19] Peter Redpath modestly complained about drawing undue attention to himself, but he was delighted to finally see "the Babylon or Museum that I have built."[20]

In his inaugural address, Sterry Hunt reflected on the differences between the Montreal meetings of 1857 and 1882. Twenty-five years earlier, scientists such as Louis Agassiz, Augustus Gould, and Joseph Henry had actively promoted the association. This "old guard" had died off, while the likes of Benjamin Silliman and James Dwight Dana (both nearly seventy) were too frail to attend.[21] Paleobotanist Leo Lesquereux declined to travel as well, citing old age and deafness, conditions which only served to further aggravate his poor command of English.[22] An even more elderly Asa Gray – one of the "old fellows," as he himself put it – dithered about whether to make the trip to Montreal and eventually decided to stay home. He remarked, "How few of us are left, after the mortality of this fatal year," on hearing that both William B. Rogers of the Massachusetts Institute of Technology (an AAAS past-president) and the association's treasurer, William S. Vaux, had died.[23] Instead of dwelling on these losses, Hunt struck a more positive note. He pointed out that this "infant university ... has under his [Dr. Dawson's] wise care attained a vigorous and flourishing manhood, and the collections in natural science, gathered with few exceptions during his administration, now suffice to fill the great museum just completed." James Hall, probably acting on cue from Dawson, used his introductory remarks to plead for teaching staff to utilize the impressive facilities.[24]

Members of the AAAS did their best to inaugurate the new building. Reverend H.C. Hovey used the modern amenities included in the lecture theatre during the presentation of his paper on "Caves and Cave Scenery," which he illustrated with lantern slides. Natural history dealers, such as the Philadelphia mineralogist A.E. Foote and the Ward & Howell taxidermy firm of Rochester, N.Y., exhibited their wares in rooms not given over to the sessions on geology, biology, and microscopy.[25]

The week-long Montreal meeting of the AAAS was the largest to date, with the exception of the Boston meeting two years earlier, which had registered sixty names more than Montreal's 937. But the wide variety of entertainments offered to participants was as important as the scientific papers (displayed in the published proceedings of the association). The local committee welcomed visitors with a "sumptuous lunch" and hosted a reception following Brush's address on the "Progress of American Mineralogy."[26] The Montreal Athletic Association and the Microscopical Club held evening soirées. The Art Association gave all AAAS members free passes to its gallery. Prominent Montreal citizens, including the Gaults, Mackays, and Molsons, held receptions at their homes and estates. Those members who wished to view the miracles of modern technology could take a harbour excursion, which showed off the Victoria Bridge and the Grand Trunk Railroad workshop. Dawson and Hunt led day outings to historic Quebec City, Ottawa, and nearby Lake Memphremagog in the Eastern Townships. Even the telephone and telegraph companies allowed members free use of their wires to send messages home. As McGill engineer J.F. Torrance exclaimed to Dawson, "Our fair city of Montreal is establishing her name for an intelligent hospitality to scientific bodies."[27]

Perhaps Dawson's only disappointment was that most scientists invited from overseas failed to make the journey. Geologists T. Rupert Jones, J.W. Judd, and Henry Hicks all sent their regrets. George Mercer Dawson wrote to his father that he had seen a number of people in England who all "seemed too busy to think of going to the meeting." Despite the favourable fares arranged with steamship lines (following the precedent established at the Albany AAAS meeting in 1856), only twenty guests came from Britain and the continent.[28] Among these was William B. Carpenter, Dawson's ally in the *Eozoön* dispute and Fullerian Professor of Physiology at the Royal Institution, who delivered an evening lecture on "Temperatures of the Deep Sea" at Queen's Hall.

THE BAAS MEETING OF 1884

The opportunity to compensate for the AAAS's one deficiency – its failure to attract European scientists – occurred immediately following the conclusion of the August meeting. In the fall of 1882, the British Association for the Advancement of Science (BAAS) announced its intention to hold its 1884 meeting in Montreal. A circular issued shortly thereafter underscored the overwhelming success of the AAAS meeting and the ample accommodation found for its nine scientific sections in the buildings of McGill University. In addition, the handbill argued, the "extensive collections contained in the Museum of the Natural History Society of Montreal, and in that of McGill University" provided materials for the study of Canadian natural history.[29] One of the British scientists who had attended the AAAS meeting, arctic explorer John Rae, explained to Dawson that his happy experience in Montreal had persuaded him to help overcome opposition to Montreal as the proposed site for the 1884 gathering.[30] Even the influential columns of *Nature* lent their support to the scheme.[31]

Many BAAS members, however, did not share Rae's enthusiasm. Perhaps they feared to confront "a howling wilderness of Arctic climate, peopled by pioneers living in log cabins and constantly armed against marauding Indians."[32] Certainly, opponents of the proposal expressed the view that there would be little serious purpose in the expedition to Canada, where they expected to find "small and uninspired audiences" with limited comprehension of scientific issues.[33] Even the governor-general of Canada, the Marquis of Lorne, held strong reservations about the proposal. Because he believed that "such a number of foolish men attend these meetings," he preferred "a delegation manageable in number and respectable in talent."[34]

Next to Dawson, perhaps the Americans were the most unequivocally enthusiastic about the Montreal meetings. They anticipated innumerable scientific benefits from bringing the prestigious foreign scientific association to their shores. The leaders of the AAAS, for instance, hoped that if their own annual meeting were planned for a convenient location and time, the BAAS members might travel south to the sessions. Dawson himself, writing a leader for the AAAS journal *Science,* argued that scheduling the two national meetings in close temporal and geographical proximity would be more advantageous than holding an international meeting. The former arrangement permitted each association to transact its entire agenda without interference. Indeed, Dawson insisted that the unique combination of British, American, and Canadian scientists meeting together in two nearby cities might attract men of science from the continent and other parts of the world.[35] The only unfortunate aspect of this decision was that those who elected to attend the AAAS in Philadelphia would then miss Montreal's *pièce de resistance,* a subsidized excursion to the Rocky Mountains.[36]

Dawson's travels through Europe and the Middle East from the fall of 1882 until the spring of 1884 (defrayed by a gift of $5000 from his friends and colleagues in Montreal), limited his involvement in the meeting arrangements. In his absence, affairs in Montreal were handled by a bewildering proliferation of local committees, some of which claimed conflicting jurisdictions and displayed competing loyalties. A general citizens' committee was formed in the fall of 1882, initiated by the Board of Trade, its goal to persuade the association to ratify the decision to meet in Montreal. Somewhat later, a newly formed executive committee, headed by Thomas White, was charged with raising a subscription fund to guarantee the undertaking against financial loss. Following the procedure established for the AAAS meeting, special committees were organized to deal with invitations, conveyance, and finance. Before the year was out, another local committee was appointed by city council to liaise with the citizens' committee. Still another local executive committee was struck during the fall of 1883 to oversee finances and general organization, again chaired by White.[37]

Despite Daniel Wilson's plea from Toronto not to "allow the British Association meeting any place in your thoughts 'til you are back in Montreal," Dawson learned from Hunt and the Harringtons that matters were not progressing smoothly. White had antagonized the scientists who sat on committees, principally McGill professors, by insisting that businessmen alone were competent to manage arrangements for the meeting. Other people objected to White's strict control and the partisanship of his faction, saying that they were running the operation "as they do the [winter] carnival, or a political engine." By December 1882, White was too busy with his own re-election campaign to be able to direct the local committee properly.[38]

In an effort to remove preparations from the arena of Montreal politics, the BAAS announced a new series of appointments and committees. Vice-presidents of scientific sections were selected to ensure an equitable distribution among Canadian universities.[39] Montrealer Thomas Cramp headed a new executive committee, labelled the "local executive committee." In addition, seven new special committees and subcommittees were formed. Their chairmen and secretaries – along with the chairman and secretary of the local executive committee, as well as the officers of the old general citizens' committee and the civic committee – constituted an executive committee with control of the purse. Merchant Hugh McLennan headed this new "citizens' executive committee" that held biweekly meetings in the chambers of Molson's Bank. With fewer than eight months left before the opening, at least eight changes were made to the executives of the special committees, besides the creation of an entirely new "economics committee."[40] (One writer quipped that he was unclear about the functions of the latter committee, since there was little evidence of economy in any part of the lavish program.)[41] As Dawson, now back in Montreal, reported to his son: "There is much confusion and imbecility in the arrangements here."⁴²

Although many of their number served on these local committees, Montrealers were encouraged to support the meeting in other ways as well. They were urged to subscribe to the guarantee fund, to contribute to a "citizens' fund for general expenses," and to become life (at $50), annual (at $10), or associate (at $5) members of the association. In order to honour Hunt's promise of "free entertainment" for at least 150 British members, Montrealers were asked to sacrifice their traditional August holidays and to quarter guests in their homes. As the circular pleaded, "Montreal ought not, at such a time, to show long rows of locked and inhospitable houses to influential guests from so great a distance."⁴³ The *Montreal Herald* and the *Gazette* insisted that no matter how modest the accommodation, foreigners would appreciate a comfortable bed and hearty breakfast. French Canadians, too, were admonished to follow the example of English-speaking inhabitants because Frenchmen would be among the visiting scientists. Once the meeting had begun, citizens were requested to display flags in honour of their guests.⁴⁴

Clara Rayleigh, who accompanied her husband in his capacity as president of the meeting, described the extraordinary hospitality of the Montreal *haute bourgeoisie* in a travel diary comprised of letters home. She spoke of Montreal houses with opulently carved mahogany interiors, the details of which were obscured by the ambient darkness maintained to keep away the heat and flies. She concluded, however, that although the town appeared very pretty, it seemed "too French and idle-looking to be impressive."⁴⁵

One may wonder what compelled Montrealers to open their homes and pocketbooks to the hordes of scientists who descended upon the city in late August. Some journalists suggested that the locals sought enlightenment from the foreigners. The London *Times* spoke of the desire of "thousands of thirsty souls" for "the delicious sensation of encyclopedic knowledge." The *Montreal Daily Star* warned that no one should expect to obtain a complete scientific education by means of a week's attendance at the lectures. That person might, however, "acquire an appetite for scientific knowledge which the meeting itself will not satisfy." Personal contact with a savant was seen as the best way to stimulate an interest in his writings.⁴⁶

Indeed, no one could dispute the intellectual calibre of the international participants at the Montreal meeting. Presidents of the London geological and mathematical societies attended, as did "many distinguished professors from Cambridge, Oxford, Dublin, Glasgow, Manchester, and other Universities." In Montreal they were joined by seventy professors from North America, including Simon Newcomb and Asa Gray.[47]

While the extent of its enduring intellectual legacy to the city may be questioned[48], the BAAS provided Montrealers with one of the best spectacles in years. Most buildings on the McGill campus received minor alterations and adjustments in order to handle the unusually large crowds. All but two of the scientific sections held their meetings on campus, which meant that twenty-nine rooms had to be "cleared of their usual furniture and prepared for temporary use."[49] The Redpath Museum, however, required no such modifications, "being exactly suited to the needs of such a congress." Members of the BAAS's geological section delivered their papers in the lecture theatre, allowing nearby museum cases to be unlocked for the curious. Just as Oxford had inaugurated its new museum with the BAAS meeting in 1860, McGill used the occasion to publicize the merits of its latest addition. As opening day approached, the city road commission cleaned and watered streets in the vicinity of the college grounds.[50]

No building on the McGill campus was deemed large enough to accommodate the crowd of 1200 that attended the formal opening of the meeting in nearby Queen's Hall. The inaugural session attracted even more people, with the citizens of Montreal raising the attendance to 2000.[51] Yet even a Saturday night lecture on comets by the royal astronomer of Ireland, Sir Robert Ball (accompanied by limelight illustrations and open to the public for a ten-cent admission fee), failed to match the success of the Thursday evening conversazione hosted by the governors of McGill in the flower-bedecked Redpath Museum.

In a repetition of the events of two years before, crowds at the museum packed the main hall almost to suffocation. It was even difficult to move in the galleries and staircase. The surplus overflowed into the lecture theatre and beyond, into the temporary passage connecting the museum with Molson Hall and the library. In the lecture theatre, experiments were performed for the curious and scientific apparatus displayed, such as a polariscope for projecting spectra on a screen and a machine to liquefy gases. The ethnologically inclined could peruse a collection of objects crafted by the Zuni Indians. At one end of the Great Hall, Governor-General (Lorne's successor) and Lady Lansdowne received guests to the orchestral strains of waltzes, polkas, and marches.[52]

A variety of other entertainments occupied the leisure hours of association members during their stay in Montreal. Prominent citizens such as Mrs John Redpath, J.H.R. Molson, and Harbour Commissioner Andrew Robertson hosted garden parties (Robertson's drew 500 guests).⁵³ Day excursions visited the Lachine Rapids and the mountain at St. Hilaire; longer outings were organized to Quebec, Ausable Chasm in New York, and Lake Memphremagog. As many had done two years earlier, 150 participants travelled by special train to Victoria Bridge and the Grand Trunk Railroad workshops. The Natural History Society, Art Association, Mechanics' Institute, Amateur Athletic Association, YMCA, and Free Thought Club generously opened their rooms and galleries to association members.

Festivities reached a peak on 3 August, the last day of the meeting. The city fire brigade mounted a virtuosic display: after driving through the McGill grounds they shot streams of water over the roofs of Redpath Museum and Molson Hall. More local colour came from a lacrosse match that pitted the "Montrealers" against the Caughnawaga Indians. That evening, 2000 inhabitants and their guests attended a reception at the Victoria Rink, hosted by the citizens' committee. Geological maps decorated the walls of the skating rink, and electrical instruments were displayed alongside tables laden with refreshments.⁵⁴

In their final appraisals, newspapers emphasized different aspects of the meeting. The great scientific event of the congress, most agreed, was the news that the platypus is oviparous – cabled from Australia and read by Oxford zoologist Henry N. Moseley to the biology section.⁵⁵ The *Manchester Weekly Times* counted 553 old members and 1215 new among the participants (making the meeting about the size of an average one held in England); 910 British members made the heavily subsidized trip overseas (the Dominion government had granted $20,000 toward fares). Another journalist credited Montreal with "breaking the ice," so that Halifax and Toronto would not seem much further away than Dublin or Belfast to an association that might soon venture to Sydney or Melbourne. The Montreal *Daily Witness* interjected the only sour note. It deplored the temporary opening of a "grog shop" on the McGill campus, despite the presence of "supposed temperance men" on the entertainment committee.⁵⁶

The chain of events culminating in the BAAS meeting prompted self-congratulation among many segments of Montreal society. The good citizens who had given so many hours to sitting on so many committees felt that the $1500 surplus, donated to McGill for scientific apparatus, provided ample testimony of their honest and careful management. Even more Montrealers had contributed to the $5000 raised from membership fees and (presumably through their taxes) to the $5000 given by City Council.⁵⁷

But Dawson, in particular, had reason to be pleased with the developments he had orchestrated. In 1893, new physics and engineering buildings opened on the McGill campus, funded by the Montreal tobacco merchant, W.C. Macdonald. What inspired Macdonald's largesse, the story goes, was his attendance at one of the BAAS lectures on physical science.[58] (In the same year, the Workman Building for mechanical engineering and the Peter Redpath Library were inaugurated). Apparently, though, apart from these bequests to McGill by patrons impressed by his achievements, Dawson's tireless efforts did not go unrewarded. It was a proud moment for him when the governor-general announced his knighthood at the opening of the BAAS meeting in Montreal. Even more rewarding was his election as president of the 1886 meeting in Birmingham. As he wrote to his son, "It might be something to be the first president from the 'colonies,' and the only man who has presided over both the American and British Associations."[59]

14 Toward International Science

Participating in the meetings of the American and British associations for the advancement of science brought an obvious deficiency to Dawson's mind: Canada possessed no national scientific organization. Even when arrangements for the AAAS meeting were just getting underway in Montreal, Dawson suggested to the governor-general, the Marquis of Lorne, that they might "discuss the feasibility of our having a Canadian Association," along the lines of the American organization. Dawson imagined, however, that Lorne, who had already instigated the establishment of a Canadian Academy of Arts, was contemplating "a distinct sort of body" with a more elitist structure than the typically peripatetic and democratic associations for the advancement of science.[1] In fact, though, Lorne (son-in-law to Queen Victoria by his marriage to her daughter, Princess Louise) worked to distinguish his governor-generalship by promoting culture in all its forms.[2]

PLANS FOR A NATIONAL LEARNED SOCIETY

By December 1881, Dawson had begun to envisage the shape of the "Canadian Association," or the "Institute" as Lorne called it. Much to his surprise, Dawson learned that Lorne favoured a more comprehensive learned society than that represented by the Royal Society of London, a form that would accommodate more French-speaking members who tended at that time to excel in literature and history. Dawson, in Lorne's view, seemed particularly well suited to organize the undertaking, given his "wide acquaintance with scientific and educational men in different parts of Canada."[3] Plans were hatched in Dawson's home in Montreal, where a provisional meeting was held at the end of the month.[4] McGill Chancellor Charles Day chided Dawson for taking on yet another task: "I see you have not yet work enough upon your shoulders, but must try to get an Academy of Arts upon them."[5]

The Montreal meeting resulted in the drafting of a trial constitution. The society was to consist of two departments, literature and science, more like the *Institut de France* than any other national model. The literary section would be equally divided between English- and French-speaking members, but no linguistic distinction was to prevail in the other department, which was further subdivided between mathematical, chemical, and physical sciences on the one hand, and geological and biological sciences on the other. A list of names was drawn up; each of the four sections was to be limited to twenty individuals, an increase over the first projected figure of sixty altogether.

In terms of a name, the French contingent disliked "Academy," but thought the use of "Institute" would produce confusion with the three Canadian institutes already in existence: the (Royal) Canadian Institute in Toronto (founded in 1849), the Institut Canadien de Québec (founded in 1848), and the Institut Canadien-Français in Ottawa. All preferred the title "Royal Society of Canada," a name that brought to mind the distinguished London-based learned society.[6]

With the initiative taken in Montreal, activity soon passed to Ottawa, and in particular into the capable hands of John George Bourinot, clerk of the House of Commons. Bourinot improved the constitution by adding some lines about "practical utility" in order to speed its way through Cabinet, and advised that a bill be drafted and passed through Parliament. This was the least expensive alternative because any fees would be refunded, unlike the case for letters patent.[7] Prime Minister John A. Macdonald and government minister Charles Tupper supported the bill to incorporate the society, which was accepted by Parliament on 6 April 1883. The bill received royal assent on 25 May 1883, the day of the first anniversary of the inaugural meeting (although the Queen had earlier approved the society's title).

The idea of a literary section – included in order to incorporate sufficiently large numbers of French-speaking members into the society – made Dawson's friend Daniel Wilson tease as he tried to draw up "a list of illustrious nobodies; the more insignificant they may be, the higher will be their delight when such honors are thrust upon [them]."[8] Wilson threatened to write to Lorne, declining to be part of this "awkward squad," "not one of whom would, in England, be thought otherwise than ridiculous in such a body."[9] He thought that the scientific section could aspire to "creditable work," as could an historical or archaeological division, but what of English literature, he badgered Dawson. "Shall we write school-boy essays or criticisms on the literature of the day; or theses on the want of literature?" he asked.[10] The problems would continue, he said, when the society looked to publish its transactions, inasmuch as a "geological or archaeological paper placed between a poem of Father Dawson and a French Ode, will be an odd publication to forward to European Savants!"[11] He hoped that Dawson might persuade Lorne to drop the whole affair, saying "As for this Canadian Academy, call it the A.S.S. or [the] Noble Order of Nobodies!"[12]

The hybrid plan conjured up by Lorne and Dawson still contained a number of unresolved difficulties on the eve of the first meeting. The government might well view with disfavour an explicitly elitist organization, after the model of the Royal Society or the French Institute. Dawson noted, in addition, the problematics of finding people of sufficiently high standing: they had had to "eke out" the English literature list by adding philosophical writers, and they could agree upon only eight geologists. Dawson also quibbled with Lorne's idea of including professional men, such as engineers. This notion presented further obstacles, inasmuch as the society would thereby tend to follow the model of a "large popular assemblage" such as the BAAS – not what Lorne had proposed originally.[13] Still, the two founders overcame their differences in the interest of preserving their joint conception.

INAUGURAL MEETING OF 25 MAY 1882

While the ever resourceful Bourinot arranged for the inaugural meeting to take place in the Parliament buildings, the ever skeptical Daniel Wilson was looking at the event with "some dread." Bourinot even managed to extract the promise of reduced railway fares for members travelling to Ottawa (as he was to do in subsequent years as well), although he still worried that the meeting would prove too expensive for most.[14] In the end, all but one member of the provisional council attended. (The most persistent problem seemed to be that the annual meeting date interfered with the conclusion of the school year or other commitments of the many members; this may explain why subsequent early meetings were plagued by poor attendance.[15])

Dawson sat as president of the provisional council that directed the May gathering. Lawyer and writer Pierre-Joseph-Olivier Chauveau, as vice-president, brought diversity of language and discipline to the executive of the fledgling society. Dawson's colleagues and cronies appeared as section heads: Sterry Hunt for mathematical, physical and chemical sciences; A.R.C. Selwyn for geological and biological sciences; and Daniel Wilson for English literature, history and allied subjects.

On the morning of 25 May, participants assembled in the Railway Committee Room of the House of Commons, where details of the constitution were presented to the membership. Although all founding members had been nominated by Lorne (obviously with the guidance of Dawson and the provisional council), the mechanisms for subsequent elections were now explained. Other details were also specified, particularly those concerning election of officers and procedures to be followed at subsequent meetings.

In the Senate chamber that afternoon, Lorne addressed the assembly of members who had made their mark via notable works of "imagination," or "the study of nature." He noted that just as America had begun to rival Germany or France in terms of its scientific expertise, Canada, too, had produced its share of individuals "whose achievements in science have more than equalled in fame the triumph of [its] statesmen."[16]

Dawson, the scientist, followed Lorne, the statesman, in addressing the group. He spoke of the "freedom and freshness of a youthful nationality" that gave the society great opportunities. He explained that a situation too youthful would bring more disadvantages than benefits. This had been the case forty years earlier, he said, when he was looking for "some means of scientific education" in Canada and had been forced to travel abroad. At that point, Canadian colleges possessed no regular curriculum for natural science, except for scattered courses in chemistry and physics. There were few natural history collections (only a handful of private cabinets), no Geological Survey of Canada, no special schools of practical science, no scientific libraries, no scientific publications, nor any public employment for aspiring naturalists.[17] Now, said Dawson, much progress had been made in Canada on all these counts. Canadian science even found itself aided by "the magnificent and costly surveys of the United States," which "freely invade Canadian territory whenever they find any profitable ground that we are not occupying." Dawson the colonist then began to speak of the "evils of isolation" that had done so much to shape his own scientific career.

Dawson's principal complaint (and here he adopted almost an autobiographical tone) related to the lack of a mechanism in Canada for the adequate publication of scientific results. He noted that the Canadian naturalist often had to content himself with the publication of his work "in an inferior style ... poorly illustrated," which automatically gave it "an aspect of inferiority" compared to work done elsewhere with its "sumptuous publication and illustration." As an "added mortification," said Dawson, he consequently found his work overlooked and neglected. While "looking in vain for means of publication," Dawson added, the fruit of his "long and diligent labour" might slip away because some publication abroad had anticipated his findings. As well, collectors in Canada tended to send their information to specialists abroad who commanded impressive means of publication unavailable to "equally competent men here." Dawson assumed, probably erroneously, that many of those present had, like himself, been victims of this unhappy chain of events.[18]

The solution for Canadian science and the reputation of Canada, continued Dawson, was for the new Royal Society of Canada (RSC) to acquire the means from the government to publish its own *Transactions*. It might be chimerical to aspire to the world-renowned publications of the Royal Society of London or the Smithsonian Institution, but certainly the RSC might emulate the example of the Academy of Natural Sciences in Philadelphia or the Boston Society of Natural History. Dawson then catalogued the beneficial results that would ensue from such a procedure. Such a Canadian journal of record, he argued, could provide its affiliated societies with the means of publishing memoirs "too bulky and expensive" to appear in their local journals.

A much less personal and more general address succeeded Dawson's; delivered by Chauveau, it represented the interests of literature and the French language. Altogether more than fifty papers were presented to the four sections (with over thirty of these later published in the Royal Society's *Transactions*). The Montreal *Gazette* characterized them as motivated by "a well-directed enthusiasm and a spirit of harmony."[19] Nonetheless some individuals were more enthusiastic than others. Only six papers were read to the English literature section (including a poem by Father Dawson). Of the fourteen papers read to the physical sciences section, four came from one member, John Bradford Cherriman. At the closing session of the meeting, on 27 May, Dawson was acclaimed as president for the following year, a task he later epitomized as "a gigantic labour encompassed with difficulties, and which occupied much time."[20]

PERSISTENT PROBLEMS OF THE ROYAL SOCIETY

In his inaugural address, Lorne had pleaded that the new society should be impeded "by no small jealousies, no carping spirit of detraction."[21] But external enmities, instigated by individuals passed over for membership, preyed upon the infant academy. The Toronto *Globe* ridiculed the organization as "a mutual admiration society of nincompoops," and inflated a personal quarrel (between Sterry Hunt and Père Hamel) into an exit by "the whole of the Quebec Clergy."[22] As one member from Quebec put it, the press "is busy besmirching them with mud," although even the BAAS had had to endure several years of such criticisms.[23] Nicholas Flood Davin, aided by H.J. Morgan, singled out Bourinot for attack in a "scurrilous pamphlet."[24] Some jealousy was defused by allowing local scientific and literary societies to affiliate with the Royal Society (as the NHSM did subsequently).

The question of financial support for the fledgling society remained, however, and Bourinot explained that the grant proposal would have to appear in the annual budget for approval by Parliament. He also suggested that members contribute a small fee, especially given that Dawson was now paying certain printing costs out of his own pocket.[25] At the second meeting of the society in 1883, the governor-general announced that an annual grant of $5000 had been approved by the House of Commons to defray the cost of the transactions. Whereas he admitted that the cost of "accurate and finely executed engraving of beautiful drawings for the illustration of scientific papers" would have to be borne, he said it was simply too costly to translate each paper into the other founding language given the limited size of the grant. He commented on the paradoxical situation that emerged, whereby the public supported the society, but could not be expected to understand the communications that emanated from it.[26] Bourinot, too, remained cautious about the extent of government largesse, warning: "I hope that there will be no slip yet between the cup and the lip. But my long experience in politics leads me to take nothing for granted."[27]

The society was constantly worried about smoothing the ruffled feathers of its French members, while working to ensure that the organization adequately represented all parts of Canada. When Alexander Murray was elected as a charter member, Dawson explained that although Newfoundland was not part of the Dominion, he hoped that his presence would amount to a "scientific annexation," at least.[28] Nevertheless, after Murray's death in 1884, the society sought not another Newfoundlander, but a representative of the Northwest and the Hudson's Bay Company.[29] Dawson vigilantly guarded against "the jealousy attracting to too many McGill and Geological Survey men among the membership."[30] He also worked to guarantee that the society would remain the preserve of scientists and literary scholars, and not be used to reward those involved in other pursuits, however worthy.[31]

On the other hand, both Dawson and Lorne took pains to defend the elitist character of the society. Dawson explained that the society was open to anyone who had produced an original work "of at least

Canadian celebrity." This implied the exclusion of people of merely "local standing"; otherwise, the society would become "a great popular assemblage" characterized more by "mere receptivity than by productiveness." Nevertheless, the society was at the same time to be exclusive and inclusive, "in that if offers its benefits to all." Dawson concluded:

Science and literature are at once among the most democratic and the most select of the institutions of society. They throw themselves freely into the struggle of the world, recognise its social grades, submit to the criticism of all, and stand or fall by the vote of the majority; but they absolutely refuse to recognize as entitled to places of importance any but those who have earned their titles for themselves.[32]

Lorne reiterated Dawson's view a year later, against those who argued for a more equitable geographical distribution among the membership. He maintained that individual eminence must be recognized as the primary consideration for membership, as demonstrated by the completion of a meritorious work. He noted that it might occasionally happen that this "personal distinction in authorship" would be "the happy possession of only one part of the country."[33] If so, those distinguished authors deserved proper recognition by a national learned society.

PEACE AT LAST

Less than a year after the inaugural meeting, most of the society's growing pains had apparently eased. Wilson made his peace with the English literature section (which included his specialities, archeology and history), and even offered to "stir up" the members in order to "make some show of work, out of courtesy to the Marquis." Indeed, within a few years, he would serve as president of the society. For his part, Lorne came to accept that the English section would likely always be the weaker of the two divisions.[34] In another development, Chauveau succeeded Dawson as president of the society. Lorne spoke of the dual advantages of this action: of gratifying the French members, while showing the English as "doing a generous thing very conducive of union in the future."[35] Bourinot was sorry to see Dawson step down, saying that his experience with societies rendered him especially "useful."[36]

The RSC's decision to affiliate with local literary and scientific societies was critical to its integration within the intellectual life of Canada, enabling it to serve as a force for cultural unity. Twelve societies sent delegates to the meeting of the RSC in 1883; by 1891, the number had doubled.37 For an organization such as the Natural History Society of Montreal, for example, the association proved to be especially beneficial. The RSC offered its serious scientists an outlet for discussion of their work and an avenue for publication. As a result, the society could concentrate its attention on science popularization, which responded more closely to the interests of most of its members. The NHSM's support for the national organization is clear from its invitation to the Royal Society of Canada to hold its annual meeting in Montreal in 1891, the first time the new society was to venture outside Ottawa.38

The centenary medal of the RSC honours Lorne, Dawson, and Chauveau. Although Lorne returned to England in 1883, he would always be remembered in Canada by "his two children": the Royal Society and the Canadian Academy of Arts.39 Unlike Lorne, Dawson continued to play an active role in the society even after his early years as founder and first president. He almost never missed an annual meeting whatever his member status.40 He advised Chauveau (his vice-president and successor as president) to heed the advice of the governor-general, who served as honorary president of the society, reminding him that he represented not just the French section, but the whole organization.41 Dawson also lobbied the government for funds, in addition to presiding over section IV, which dealt with geology and allied sciences. Moreover, he constantly solicited papers for the society's proceedings from his network of correspondents across Canada. The *Gazette* characterized him as "the right man in the right place," an individual whose patience, tact, and reasonableness never failed.42 The extent of his commitment is summed up in his wife Margaret's carping: "I shall rejoice to hear of the meeting of the Royal Society at Ottawa being over. I look upon these as useless wastes of strength and time and hate them accordingly."43

DAWSON'S BAAS PRESIDENCY: BIRMINGHAM, 1886

There appeared to be no limit to Dawson's organizational talents. His extraordinary ability to placate various individuals and interest groups, as well as his uncanny political sense, were clearly demonstrated in the smooth functioning of the Montreal association for the advancement of science meetings and the launching of the Royal Society of Canada. Although nothing marred his execution of these tasks in the early 1880s, the situation was to change shortly thereafter. Just six months after the BAAS meeting had met in Montreal, the first signs of a new epidemic of smallpox began to appear. Anti-vaccination riots rocked the tranquillity of Montreal; enrolment at

McGill plummeted; and "all trade and enterprise" were driven from the city. Dawson faulted the police for "cowardice and mismanagement," as well as the French who, through prejudice and ignorance, allowed hundreds of their children to die.[44]

Dawson's knighthood, awarded during the 1884 BAAS Montreal meeting, lauded his ability to boost Montreal and Canada's scientific profile. Dawson later reflected that, as a man of science, he had escaped the "somewhat insidious comments in the press," that often accompany the conferring of knighthoods on colonists.[45] A further indication of the high regard in which he was held was his nomination for the presidency of the BAAS meeting in Birmingham in 1886, with its "additional testimony to the success of the Canadian meeting."[46] Unlike "the mean contempt for colonists" displayed in Britain's political press, noted Dawson, British scientists seemed to receive colonials as their equals.[47] Writing privately to his son, Dawson complained of the expense and time involved in making the trip to England; nevertheless, he admitted that because the Birmingham meeting was likely to be larger than the next one in Aberdeen, "the kudos [are] great." He also reflected that "it might be something to be the first president from the 'colonies,' and the only man who had presided over *both* the *American* and *British Asssociations*."[48]

In 1886, the year of the Colonial Exhibition – the first exhibition held in Britain devoted to displaying the products of the colonies – Dawson probably seemed to be the obvious choice to preside over the BAAS meeting.[49] Indeed, he had been involved earlier with the Fisheries Exhibition (held at South Kensington in 1883) and in ensuring proper representation for Canada at the international expositions held periodically since the middle of the century. Dawson urged Sir Charles Tupper, charged with Canada's representation at the Colonial Exhibition, to reflect the scientific work of the Dominion in an appropriate way. Complaining that mistakes made by the Nova Scotian naturalist, David Honeyman, had incurred ridicule for Canada at the Fisheries Exhibition, he spoke of his trust that the Geological Survey staff would redeem Canada's reputation, putting the country on a par with Australia and India, who will "try to make a mark in scientific matters." This Colonial Exhibition was especially important, he said, because the products of the colonies would be given pride of place, unlike earlier expositions where they were overshadowed by older and "more advanced" countries.[50] Dawson explained that his Birmingham presidency made him especially concerned with this issue.[51]

Because Dawson believed that this might well be his last trip to the "old country," he turned the voyage into a triumphal, regal farewell tour; one correspondent called it his "campaign in England."[52] He and Margaret spent some time in London, before travelling on to Birmingham the week before the meeting, where they stayed at the home of the mayor, Thomas Martineau.[53] They managed to visit old friends on the way, including William B. Carpenter and his wife, in Manchester. Other long-time acquaintances and correspondents arranged to see him in Birmingham, such as the Redpaths and J.R. Pattison, the Religious Tract Society's editor.

The town of Birmingham spared no effort to make this the most successful BAAS meeting ever held there. From the fronts of the galleries in the Town Hall were suspended "bannerets" bearing the shields of past presidents. Dawson's flag had been confected by the Birmingham firm of Legg & Co. It sported three crows, or daws, spaced across a golden diagonal band on an azure shield. Above the shield danced a six-legged starfish; inscribed below was the Rankine family motto *Fortitier et Recte* ("boldly and justly"), which replaced the Dawson motto *Toujours propice*.[54]

Margaret fretted about her husband's address, since Williamson had confidentially warned her that any mention of theology or the Bible would be a "fatal mistake." He explained that Dawson "stood *absolutely* alone" among scientists in his vigorous defense of revelation.[55] Margaret need not have worried; her husband's presidential address on the geological history of the Atlantic Ocean (an appropriate topic from one whose life was both linked to, and separated from, British science by this vast expanse) won him general acclaim.[56] One newspaper likened it to "a fairy tale of science, sounding the Atlantic Ocean to its lowest depths, and ascending to the highest peaks of the loftiest mountain ranges."[57] A spate of violent earthquakes in the United States and Europe made Dawson's words particularly timely; as the Marquis of Lorne wrote, they acted like "a gigantic natural chorus" to his speech.[58]

Dawson's oratorical skills reminded one reviewer of Lyell, in terms of his "cautious reluctance to accept new doctrines, combined with a conscientious outspokenness in proclaiming his convictions when fully matured." Dawson was said, too, to have appropriated Lyell's "certain simple felicity of illustration."[59] Another found in Dawson's talk "the simple strength of diction" of which Darwin was "the unconscious master." Dawson, he added, seemed more concise than Herbert Spencer but not as brilliant as Huxley.[60]

Dawson's firm religious convictions could not fail to impress his auditors. In his introductory remarks concerning the new era in English science associated with the names of Darwin, Wallace, and Spencer, Dawson still insisted that "neither observation, experiment, nor induction" could fully solve the mysterious question of origins.[61] The

Manchester Guardian believed that the talk would be remembered as a "counterblast" to John Tyndall's famous Belfast address of 1874.[62] The Irish geologist Edward Hull praised Dawson for delivering a talk "different from the irreligious addresses" of recent years.[63] Even one of the most nonreligious of British scientists, Thomas Henry Huxley, sent congratulations, assuring Dawson that his own presidency at Liverpool in 1870 had heralded one of the hardest week's work in his life.[64]

Only the English edition of *Figaro* complained that the Birmingham proceedings were "exceptionally dull," with not a single paper of note.[65] More typical of the general response to the event was that of Mayor Martineau, who assured Dawson that local residents and visitors alike were pleased with the meeting.[66] Dawson's London publisher, the firm of Hodder & Stoughton, spoke of its appreciation of the publicity generated by the British tour, while Cambridge University announced that Dawson had been awarded an honorary degree.[67]

One correspondent urged Dawson to see the presidency at Birmingham, not as a "climax" to his life, but as the end of one era and the beginning of another.[68] Certainly, Dawson began to view with greater satisfaction his role as a "colonist" and to explore ways in which he might facilitate communication among scientists of the British Empire. On his return to Montreal, he urged Canada to send a scientific delegation to the Sydney Centennial in 1886.[69]

TOWARD INTERNATIONAL GEOLOGICAL UNION

Ever since its early days of organizing into scientific groups, the North American scientific community had displayed a bias toward geological pursuits. Ten of the first forty presidents of the American Association for the Advancement of Science were geologists.[70] By 1888, estimates showed that around 200 geologists were at work in North America.[71] Beginning in the early 1880s, these geologists began to express dissatisfaction with the AAAS. Dissident rumps, led by C.H. Hitchcock and the Winchell brothers, Alexander and Newton, formed at the annual meetings, including that held in Montreal in 1882. The geologists objected that these summer meetings interfered with their fieldwork, were too broad in scope, and were more social than scientific. At the Montreal meeting, they conceived the idea of a new journal to represent their interests, the *American Geologist*. The journal was not launched until 1888, however, the year the Geological Society of America was founded.[72]

Despite the fact that his close friend James Hall was named as first president, Dawson at first declined to join the society.[73] In 1889, New

York University Professor J.J. Stevenson entreated him (and George Mercer Dawson) to join, explaining that "if the elder men whom every young man knows are missing, the young men will not come in."[74] Dawson had no objection to adding his name, he told George, but he could not find the time to attend the meetings. For his part, George took offence at the moniker, quipping that "whenever they want our help they associate us with 'American geologists,' but when they wish to say we are wrong then we are 'Canadian geologists.'" The fact that the new society was to meet with the AAAS in Toronto the following September seems to have persuaded the Dawsons to join.[75]

Another thrust of the dissident geologists within the AAAS was to try to effect their platform within an international context. As a member of the founding committee, Dawson actively promoted this undertaking, initially called the International Congress of Geologists (later, the International Geological Congress).[76] The idea of the congress had been hatched as early as 1876 as a means of breaking down national barriers in geology, especially through the promotion of common conventions in nomenclature and map coloring.[77] During Dawson's lifetime, triennial meetings were held in Paris (1878), Bologna (1881), Berlin (1885; postponed from 1884, due to an outbreak of cholera in southern Europe), London (1888), and Washington, D.C. (1891).[78] Hundreds of participants converged on these meetings, citizens of the host country typically dominating the attendance figures.[79] Both organization and procedures were "common-sense adaptations" from other congresses, allowing one week for scientific sessions (with publication of proceedings) and field excursions, both before and after.[80]

The first congress meeting was timed to coincide with the Great Exhibition, held in Paris in 1878. This was the largest of the international expositions to date, with the British colonies (past and present) occupying nearly one-third of the space reserved for countries outside France.[81] Dawson's contributions were given a place of honour at these affairs: a set of *Eozoön* specimens prepared by the Geological Survey of Canada, for example, illustrated a copy of his *Dawn of Life* (as had been the case two years earlier at the Philadelphia exposition).[82] Both Dawson and his son George sent papers to the Italian meeting of 1881.

Five years later, however, Dawson had changed his opinion of these gatherings; he sent a letter of resignation to the American geologist, Persifor Frazer.[83] Apparently, Dawson objected to the actions of the American committee of the Geological Congress following the Berlin meeting of 1885. He called the Americans "a quantity of geological cranks" because of their inability to communicate adequately with their European brethren and their failure to resolve questions of nomenclature. Nonetheless, he condemned Fraser for circulating his letter, rather than the official resignation, in which the offensive phrase did not appear.[84] James Hall later tried, without success, to patch up the matter.[85]

IMPERIAL SCIENTIFIC FEDERATION

A Swiss member thereupon suggested that Dawson form a Canadian committee of the Geological Congress. Dawson, however, viewed the idea with disinterest.[86] From this point on, Dawson, disenchanted with the congress, turned his keen organizational talents elsewhere. He explored the prospect of bringing geologists from throughout the British Empire together, under the wing of the British Association and the Royal Society. (As early as 1857, Dawson had proposed a common meeting on "neutral" Canadian territory, if not a general unification, of the British and American associations for the advancement of science.)[87] Successive Royal Society presidents T.H. Huxley and George Stokes expressed support for the idea of a scientific federation. Dawson had been inspired, as well, by the Colonial and Indian Exhibition (attended on his way to Birmingham in 1886), with its display of Britain's and dependencies' geological riches and resources. Such a scientific federation, thought Dawson, would fittingly commemorate the fiftieth year of Victoria's reign, in 1887.[88]

As Dawson explained to Royal Society President Stokes and others, although "different languages and habits of thought" had impeded unification within the International Congress of Geologists, this was not to be the case for the British Empire. Dawson foresaw a "more limited union," originating from the impetus of the British Empire and later joined by all English-speaking countries, perhaps ultimately bringing about a wider participation. Dawson maintained that within the empire, and in sciences generally, geology would lend itself to union "more readily" than other areas of knowledge. He went on to describe Britain's controlling interest in world geology as a result of the extension of British stratigraphy by colonial geologists in Canada, India, Australasia, South Africa, and even in countries "not under the British flag."[89]

Many in Britain applauded Dawson's initiative. The *Manchester Examiner*, crediting him with transforming the proposal from its earlier "nebulous shape," said that although the universalism of science should recognize no boundaries and worldwide scientific union appeared unattainable, perhaps imperial union could provide the next best thing.[90] *Nature* supported Dawson's suggestion in a leader on "scientific federation." Certainly those with colonial experience welcomed Dawson's proposal. GSC director A.R.C. Selwyn agreed that the Geological Congress would never produce "any definite result," although that might have been otherwise if it had confined its attention to the maps and reports published by geological surveys.[91] The Marquis of Lorne praised the attempt to unify individuals in "non-political ways."[92] The Royal Society of New South Wales pledged its cooperation and offered to bring the proposal to the attention of the delegates of Australian and New Zealand scientific societies, convened for the first time ever in March 1888.[93]

Despite the support of such influential British scientists as Cambridge professor Thomas McKenney Hughes and Irish geologist Edward Hull, both British scientific societies came to view the proposal as chimerical. In 1886 (during Dawson's presidency of the general association), the British Association Council not only rejected Sydney's invitation to hold their 1888 meeting there, but also declined to support Dawson's ideas about a scientific federation. Dawson, who had hoped to connect his presidency with these resolutions, protested by refusing to attend the 1887 meeting in Manchester.[94] The Royal Society asked Dawson to describe "the proposed practical steps" by which the union could be effected.[95] He did so in detail, using a conference convened by the Royal Society as the first step.[96] Ultimately and predictably, however, the Royal Society declined to lend their support to Dawson's proposal, seeing it as a potential source of conflict with the Geological Congress.[97]

Clearly, Dawson's dreams for international scientific cooperation ran into the reality of vested scientific interests at home and, especially, abroad. His own nation was too young to act as a cohesive force or to stimulate other countries to join in its organizational efforts. Dawson's personal commitment to internationalism developed as a natural extension of his network of correspondence, his leading role in a range of scientific institutions, and his talents as an administrator.[98] But his efforts proved to be premature, largely because of the absence of any widespread recognition of Canada's emerging importance as a scientific power and world actor.

15 No More Toil

> Blessed are the dead which die in the Lord that they may rest from their labours and their works do follow them.[1]

The Montreal *Gazette* described Dawson's funeral, held on 21 November 1899, as "one of the most numerously attended Montreal has ever seen." The old library of the central building on McGill campus was transformed into a mortuary, with the pastor of the Stanley Street Presbyterian Church, where Dawson had been an active parishioner, presiding over the service. The hearse that carried the coffin to the Dawson plot in Mount Royal Cemetery threaded its way through the university grounds, subsequently proceeding down avenues lined with undergraduates paying their last respects.[2] After almost half a century of good works in McGill and Montreal, Dawson came to rest not in his native Nova Scotian town by the sea, but in his adoptive city on the hill.

In the eulogies delivered by colleagues and friends, Dawson was portrayed as an administrator of the stature of Charles William Eliot at Harvard or Daniel Coit Gilman at Johns Hopkins. It was said by some that Dawson had made McGill into "the Canadian Oxford," and that its prestige was excelled in America only by Harvard.[3] Others pointed out that he had transformed McGill virtually singlehandedly from a "tiny, poverty-stricken provincial school" into "a well-endowed university of worldwide reputation."[4] The overriding sentiment of the memorialists might be expressed in a single sentence: "McGill was Sir William Dawson, and Sir William Dawson was McGill."[5]

Even six years earlier, as Dawson stepped down after thirty-eight years as principal of McGill in 1893 and moved to a more modest residence on University Street, it appeared that no part of the university's activities had escaped the impress of his guiding hand. Student enrolment, including by then women as well as men, totalled nearly 1000, with teaching staff numbering more than one hundred. The older faculties of law and medicine had been reinvigorated by handsome endowments and buildings and joined by new faculties, such as applied science.[6] Hugh McLennan, perhaps better than any other writer, has eloquently captured the significance of the Dawson era at McGill. He portrays these years in terms of the growth of famous families, with Dawson cast in the role of the paterfamilias:

In the next generation there appears a family head who is almost a tribal leader. He has the immense inner self-confidence of a limited man of genius. His mind moves tenaciously along a single groove, but because that groove leads forward, and because his energy is enormous, he cannot possibly fail. He expands the family resources, he rules it like a patriarch, he sets it into a mould with many deficiencies and not a few ungainly protuberances. Viewed from one standpoint he is ridiculous, viewed from another he is sublime. He is wrong as only a pig-headed man can be about the value of certain things which the best minds and spirits of all ages have considered priceless, yet he is wrong for reasons natural to his own limited nobility, and about one other thing he is never wrong at all: if a family serves a community to the best of its power, even the most indifferent of communities in the end is sure to support the family. After a life of struggle in which the real gains were immense but the show of them small, the patriarch, now an old and tired man, at last has time to pause and look around him. His work is done, his family is secure, and before it lies a road transcending his imagination.[7]

The momentum gained in Dawson's zealous pursuit of McGill's scientific fortunes carried other institutions along in its wake. His success in bringing outstanding scientific talent to McGill, including the botanist David Penhallow and the young medical doctor William Osler, reaped a double benefit: these individuals also became active members of other institutions in Montreal, such as the Natural History Society of Montreal, where their brilliant research brought renown. McGill's remarkable scientific facilities likewise played a significant role in attracting the American and British associations for the advancement of science to Montreal during the early 1880s.

Dawson's work as an educator outside the boundaries of McGill involved the advancement of educational enlightenment at all levels. The superintendent of Quebec's Department of Public Instruction, Boucher de La Brière, eulogized his efforts: "Although Sir William spoke a tongue and professed a religion other than my own, ... his achievements in the world of science and his work as an educator reflect credit upon all Canada."[8] His sound sense of administration not only advanced the cause of English Protestant education in Quebec, but helped to chart a clear course for the foundering Natural History Society of Montreal, which, as its fortunes improved, was to assume an active role in popular education via its museum and annual conversazione.

One newspaper of the day claimed that Dawson would have been famous as an educator, even if he had "never discovered a fossil, or made a geological map."[9] Yet he had come to his life's enduring work as an educator almost by default, as a result of the lack of opportunities available to him as a scientist, whether that be the chair in natural history denied him in Edinburgh or the absence of professional employment as a geologist in Canada. His inaugural address at McGill made clear his awkwardness filling the large shoes of the "educationist." At that time, despite his pioneering educational work in the Maritimes, he felt inadequate to the new responsibilities thrust upon him, given his youth and inexperience in the realm of higher education.[10]

One might argue, equally forcefully, that Dawson would have been famous as a scientist, even if he had never served as principal of McGill. His painstaking research in the geology and paleontology of coal formations alone is enough to guarantee him an important position in the annals of the history of science. But severe damage to his scientific reputation resulted from his having been seen as the last survivor of that "pre-Darwinian group of naturalists headed by the elder Agassiz."[11] His scientific work became discredited not only because of his outspoken opposition to Darwinian evolution, but also because of his stubborn adherence to increasingly precarious positions in disputes over the nature of *Eozoön* (seen by his opponents as "fanatical and baseless") and the efficacy of glacial action (his detractors denounced his "heretical" views).[12] As he remarked to the sympathetic James Hall, who well knew the difficulties inherent in deciphering the fossil record, "I have been more careful than most men in the distinction between plants and animals; but mistakes will occur."[13] Dawson expressed his eagerness to "gather together" into a comprehensive paper "all the mistakes I have made in referring fossils to wrong groups, which are many," but he wished at the same time to put them against "the things I have helped to put right."[14]

If Dawson found himself often on the unpopular and ultimately unsuccessful side of scientific debates, he adopted a view of the history of science that gave him solace. He saw scientific advance as a series of oscillations dictated by adherence to different theories. Individuals tended to be seduced by one-sided "very decided doctrines," especially by the most fashionable and newest scientific theories of the day. Unlike those many scientists who tended to accept the fad of the moment, Dawson saw himself as a moderate, a skeptic of these "swing[s] to the extreme with every bob of the scientific pendulum." Moreover, he had been caught up in the process of scientific specialization and fragmentation. Not only had his work been misrepresented by specialists, but his own specialized contributions – particularly his investigations of the Devonian fossil flora – had been misapprehended by nonspecialists.[15]

One historian has argued that Dawson's "religious fundamentalism" and "closed system of thought" constituted "a reaction to his fear and dislike of ultramontanism and his sense of cultural and intellectual isolation in French Canada."[16] Such a view, with its assessment of Dawson's personality as narrow and negative, violates my own sense of the man. More characteristic was Dawson's positive response, his acceptance of personal difficulties as a spur to greater efforts, and his channeling of his energies into other realms where he might accomplish more. Rather than accepting isolation, Dawson moved vigorously to promote Canada's scientific fortunes both in North America and at an international level. The result was a range of scientific institutions, with the Royal Society of Canada at the pinnacle, that would serve the country well as it moved into the twentieth century.

Toward the end of his life, scientific honours and distinctions indicated the respectability that Dawson had achieved in official circles. In 1884, he was awarded an honorary LL.D. from the University of Edinburgh; by the mid-1890s, he was elected a corresponding member of the Geological Society of France and accorded the rare distinction of honorary fellowship in the Royal Society.

Despite these accolades and the powerful legacy of Dawson's range of achievements in Montreal, his family worried over whether history would remember the importance of his contributions. Anna fretted over her mother's choice of wording on the tombstone, preferring "Jno. William" or "J. William" in order "to give importance to the name."[17] But the concern with posterity's evaluation emerged most fully in the furor that developed in the family circle over the publication of Dawson's autobiography, *Fifty Years of Work in Canada*.

"THIS WRETCHED BIOGRAPHY BUSINESS"[18]

Just a few months before Dawson's death, as he lay weakened by stroke and bouts of pneumonia, son Rankine precipitously returned to Montreal. As the prodigal son, Rankine spent two months at his father's bedside, providing a great source of comfort and support to both parents. After his father's death, Rankine returned to England to fetch his wife and baby daughter, intending to settle in Montreal and establish a practice as a physician.

Perhaps when death had removed the long shadow of his father's *persona,* Rankine could for the first time contemplate a return to his birthplace. But a permanent move was never to occur. The intense discord that erupted in the Dawson family over his involvement with publishing *Fifty Years of Work in Canada* would dictate otherwise. Rankine assumed the editorship of the book and published it quickly – "impetuously" according to his mother. He, however, interpreted these responsibilities as his "sacred engagement" to his dying father, even over the protests of his siblings.[19]

In the family's estimation, George was the most like "Papa" in temperament and application; more than anyone else, he had been capable of entering into his father's thoughts and feelings.[20] In the view of the family, then, he was the obvious choice to edit his father's memoirs. Even Rankine at first deferred to George as editor of the autobiographical manuscript, pledging his assistance should it be useful.[21] However, George found that his new responsibilities as survey director meant that he could not continue the undertaking. Anna later surmised that Rankine's involvement had been dictated by their father in his final days only because he feared that the task would kill George.[22]

Dawson's widow, Margaret, seems to have been instrumental in placing the manuscript in Rankine's hands; having detected that Rankine was becoming "despondent," she may have believed that the responsibility would do him good. (Later Anna suggested that it was her mother's pressure that had persuaded "Papa" to allow Rankine to undertake the job in the first place.[23]) Dawson's widow nevertheless confided that she would "unhesitatingly" prefer George to do the job and felt that Anna might better portray the "home-life," but that neither seemed to have the necessary time available to execute the task.[24]

Just a few months later, Anna reached the conclusion that her mother had made a terrible mistake and expressed her surprise that Rankine had decided to undertake the job. She felt that, of all the children, Rankine had sympathized with, and understood, their father the least. Even as an adult, he had constantly criticized his father's actions, conduct, and whole line of work.[25] Anna and George, fearing that Rankine's lack of empathy would seriously damage the end result, began to ponder how the project might be wrested from his control. From a distance, William and Eva offered quiet support for their siblings' antipathy to Rankine. However, George, who was to hold the greatest respect for William as always "very fair and sound in his opinions and views," called Eva's "pretend-you-don't-know attitude" absurd in this context.[26]

As events unfolded, George and Anna began to dwell on Rankine's character flaws. George confided that he questioned Rankine's sanity;

altogether he found him "inaccessible" to argument, although he did admit that the two of them were "of a peculiar (or of two peculiar) compositions."[27] Anna, too, referred to Rankine's "nervous sufferings" and suggested that "he has a singular way of looking at things which is convincing to himself but leaves out weighty matters." Although kind and thoughtful when leading the dependent, she judged him "very difficult" at other times. His "very positive manner" seemed calculated to conceal his irresolution, while he appeared to possess "some kind of squinting vision" that prevented him from seeing things as others did.[28] Eva concurred with Anna and George that Rankine seemed rude and one-sided, altogether ignoring the feelings of others.[29] But now, compounding these deficiencies, he had placed himself in the "untenable and false position" of dictating to the rest of the family, acting as if he were their father's "sole faithful executor." George was outraged that Rankine should pretend that *père* Dawson's career and life were "in some way his perquisite and his peculiar keeping."[30]

Even though Anna maintained that Rankine did not possess "evil intentions" but only "wrong views," mercenary motives did not escape his siblings' consideration. Anna thought that Rankine had perhaps sold the biography for "a lump sum," while George doubted that any English publisher would pay for such a work.[31] Margaret, however, dismissed these petty remarks, maintaining that the $300 authorized by Dawson for the editor of his autobiography, at the time thought to be George, would barely cover Rankine's travel expenses from England. She did admit, however, that Rankine seemed to overestimate the extent of potential sales, while underestimating the cost and trouble involved in ensuring even these.[32] Anna and George also distrusted Rankine's motives in deciding to move to Montreal, where he looked to find "every advantage to brace him up, and help him on."[33]

The ostensible source of friction between George, Anna, and Rankine seemed to be their differing conceptions of what the work should be. Rankine maintained that their father had viewed the manuscript as his autobiography – not as source material for someone else's biography – to be published in its final form with only the most minor editorial alterations. As such, it could stand as a "personal reminiscence" of a long and active life. That it seemed "self-centred and egotistical" simply meant that it shared common currency with all works of the genre.[34] George and Anna argued against this view, however, maintaining that the significance of Dawson's legacy would be thereby impaired. Anna actually found her father's account too modest: no record of his extraordinary kindnesses, for example, appeared; nor did he claim any credit "except for earnest toil." The two of them believed that the manuscript should be delivered into the hands of "some outside literary man," who might supply both objectivity and embellishment in the form of anecdote and detail.[35]

The underlying problem was not so simple, however. Rankine grounded his argument in an estimation of his father's character that his siblings found exceptional and untenable. He maintained that unlike professional men who live for their clients and patients, thereby intentionally effacing themselves, their father had taken little interest in the lives of others. Rankine viewed the manuscript as the final "rounding off," the completion – even celebration – of his father's life's work.[36] In Rankine's opinion, Dawson's singular character had expressed itself through good works, not through interpersonal relationships with others.

Anna and George, in contrast, countered that their father's autobiography supplied a damagingly incomplete and misleading portrait of "Papa." Anna thought that her father's mind had by that time "lost the elasticity and vigour" necessary for supplying "a just estimate of his own achievements." She even noted that his feeling of haste was a symptom of his physical state, and had been entirely misunderstood by Rankine.[37] George contended that if this was what his father had written in his later days, it became all the more necessary to edit the work in order "to show what he was at his best and how his work was done."[38]

At the most benign level, the two siblings found the account dull and lacking in "literary charm," enlivening detail, and anecdote, little more than a "bald narration of facts," and altogether "devoid of incident and human interest." More substantive objections, however, led to their conviction that the autobiographical account was so seriously flawed as almost to be a caricature of their father. The artistic Anna found that her father had inadequately portrayed the beauty of his own life – its colour, form, and atmosphere – so as to be merely a record "like a botanical description of a fragrant flower, which one would prefer recorded by an artist's luminous and tender touches."[39]

As Anna and George continued to debate the issue, Rankine went ahead and made arrangements to publish the work, a procedure that George referred to as "highway robbery."[40] Indeed, George was so outraged by this turn of events that he considered taking legal action against his brother. The two siblings also thought to recruit William, who always seemed so "eminently just, and clear ... entirely without self-seeking or even self-importance" to speak to Rankine, but Anna feared a "lack of tact."[41] Even Eva, entirely unaware of the whole affair, wrote that it was good for Rankine to be working on something, and that he might derive some useful lessons from reading about his father's self-sacrificing life. She also managed to persuade her siblings that although the work was incomplete and imperfect, it was not unsuitable, and that it would be best to forget the matter in the interest of honouring their father and comforting their mother.[42] Anna, however, who could not even bring herself to read the book, reproached Rankine for including a portrait of her father in which he looked feeble.[43]

In the end, Rankine decided to remain in England, where he eventually died after several bouts of mental illness. The whole sad affair resulted in bitterness and a shattering of the "peace and harmony" that Margaret Dawson so desperately wanted. It provided a "melancholy close" to the life of her husband, who above all else had sought peace.[44] Anna concluded that the problem, at heart, could be resolved with the simple explanation that Rankine's plans and "modes of thought" were "on rather different lines from ours." The unfortunate tendency of most members of the Dawson family was to criticize others who did not share their ways.[45] In this context, Rankine was a maverick whose unorthodox behaviour could not be tolerated. But perhaps the words of George Mercer Dawson best sum up the legacy of the elder Dawson: "No one of his children is likely to achieve anything comparable with what he did, but we are trustees of his fame and record, and responsible for any unworthy presentment that may be made of them." Surely, he said, all the children could agree on this point and quarrel, if they must, about "our own comparatively insignificant affairs."[46]

The Dawson children lamented their father's lack of personal reflection on his life, his reluctance to claim "any credit ... except for earnest toil," as the title, *Fifty Years of Work in Canada,* so well made clear.[47] But such would seem to have been Dawson's own evaluation of his life. Always he was aware of how little time he had to indulge in any one interest. As a colonist, he believed, he was "called on to play many parts."[48] In the end, his faith never wavered, his hope never dimmed, and his good works endure to this day.

Notes

A NOTE ON SOURCES

Thanks to the generosity of Dawson's granddaughter, Lois Winslow-Spragge, who donated extensive documents to the archives during the early 1970s, the McGill Archives holds a virtual embarrassment of riches concerning the life of John William Dawson. The copious amounts of Dawson materials touch every aspect of his full and varied life. Among these materials are some 5000 items of scientific correspondence, in addition to letters that deal with the administration of McGill and family matters. (See my *Index to the Scientific Correspondence of John William Dawson* [BSHS Monograph 7] [Stanford in the Vale: British Society for the History of Science 1992].) Another twenty boxes hold papers and scrapbooks concerning science, religion, mining, travel, and so forth (McGill Archives has divided the materials into containers according to various subject headings), in both printed and manuscript versions. (See the *Guide to the Archives of McGill University*, Marcel Caya, ed., [Montreal: McGill University Archives 1985], vol. 3: 21, for a more detailed description of these papers.) Still other boxes of Dawson materials at the archives have never been integrated into the chronological runs of letters or categorized boxes. These include, for example, the Redpath Museum correspondence and minute books.

Besides the papers of Dawson himself, the McGill Archives also serves as a repository for documents from other members of the Dawson family. Among those which have proved particularly invaluable in this context are the papers of William's father, James Dawson (in three containers, arranged chronologically). Equally significant are the letters that Dawson exchanged with his wife, Margaret Mercer, beginning with their courtship in 1841. As well, a microfilm copy of a biographical memoir of Dawson and his wife, attributed to granddaughter Clare Harrington, provides some remarkable insights and family photographs. The fragmentary manuscript of Dawson's autobiography, as well as the family correspondence that its publication evoked, has proved to be especially interesting.

The Blacker-Wood Library at McGill holds the incomplete minute and record books of the Natural History Society of Montreal (NHSM). These should be examined in conjunction with the printed proceedings published in the *Canadian Naturalist* (subsequently, the *Canadian Record of Science*), a complete run of which the Blacker-Wood Library also possesses. As well, the Blacker-Wood Library has two boxes and two scrapbooks dealing with the British Association meeting of 1884; these were assembled by the NHSM.

McGill's Rare Book Department has a large collection of Dawson's books and articles, as well as a box of obituaries and biographical articles. The department also holds thirty boxes of scientific reprints presumably sent to Dawson by his correspondents, although I believe this likely represents but a small sample of the articles he received during his lifetime.

The list of abbreviations preceding "Notes" introduces the diverse archives – scattered across Canada, the United States, and the United Kingdom – that have Dawson materials. Several merit mention here for providing especially rich documentation concerning a particular phase of Dawson's life. An understanding of his education and early career in Nova Scotia is enhanced by the holdings of the Public Archives of Nova Scotia in Halifax, especially the papers of the Pictou Academy. Dawson's assault on the Edinburgh natural history chair in 1854 could only have been reconstructed via the papers of the Lord Advocate's Department at the Scottish Record Office (Edinburgh) and through the minutes of the Edinburgh Town Council (City Archives). Similarly, the archives of the Royal Society in London provide a necessary perspective on the rejection of Dawson's Bakerian Lecture on Devonian plants. For all other aspects of Dawson's life, the holdings at McGill University – at the Archives, Blacker-Wood Library, and the Rare Book Department – are unequalled.

The most useful secondary treatments of Dawson include his own autobiography, edited by his son Rankine, *Fifty Years of Work in Canada, Scientific and Educational* (London and Edinburgh: Ballantyne, Hanson & Co., 1901) and Charles F. O'Brien's volume, *Sir William Dawson: A Life in Science and Religion* (Philadelphia: American Philosophical Society, 1971) (see Chapter 1 for a discussion of these and other secondary sources). Peter Eakins supplies an interesting treatment of Dawson in the *Dictionary of Canadian Biography* (Toronto: University of Toronto Press 1990), vol. 12: 230–37. He refers there to a bibliography of Dawson's publications, which he has deposited with the McGill Archives. I have not seen this document that presumably supplements H.M. Ami's bibliography in "A Brief Biographical Sketch of Sir John William Dawson," *American Geologist*, 26 (1900), 1–57, which is marred by errors and omissions.

ABBREVIATIONS

The following abbreviations are used in the notes. Letters for which no other location is given are found in the collection of the John William Dawson papers, McGill University Archives.

- AAAS American Association for the Advancement of Science
- APS American Philosophical Society, Philadelphia
- BAAS British Association for the Advancement of Science
- BMS Boston Museum of Science
- CN *Canadian Naturalist and Geologist*
- CRS *Canadian Record of Science*
- CUL Darwin papers, Cambridge University Library
- DCB *Dictionary of Canadian Biography*
- DNB *Dictionary of National Biography*
- DSB *Dictionary of Scientific Biography*
- ETC Town Council Records (Edinburgh City Archives)
- EUL Edinburgh University Library
- GS Geological Society, London (Roderick Murchison papers)
- HUA Harvard University Archives
- HUG Harvard University, Gray Herbarium
- HUM Harvard University, Museum of Comparative Zoology
- ICS Imperial College of Science and Technology, London (T.H. Huxley papers)
- JWD John William Dawson
- MC McGill Archives
- MBW McGill University, Blacker-Wood Library
- MRB McGill University, Rare Book Department
- NA National Archives of Canada (J.A. MacDonald papers)
- NHM British Museum (Natural History), Library
- NHSM Natural History Society of Montreal
- NYL New York State Library, Albany
- OUM Oxford University, Geological Museum (Phillips papers)
- PANS Public Archives of Nova Scotia
- QFHSA Quebec Federation of Home and School Associations
- QJGS *Quarterly Journal of the Geological Society*
- RS Archives of the Royal Society, London
- SIA Smithsonian Institution Archives
- SRO Scottish Record Office (Lord Advocate department's papers)
- YUL Yale University Library

CHAPTER ONE

1 Rankine Dawson, ed., *Fifty Years of Work in Canada, Scientific and Educational* (London and Edinburgh: Ballantyne, Hanson & Co. 1901), 173.
2 For an excellent short biography of Dawson, see entry by Peter R. Eakins and Jean Sinnamon Eakins in *Dictionary of Canadian Biography* (Toronto: University of Toronto Press 1990), vol. 12: 230–7.
3 As Adrian Desmond and James Moore describe previous biographies of Darwin, in their *Darwin* (New York: Warner Books 1991), xviii. See my "New Directions for Scientific Biography: The Case of Sir William Dawson," *History of Science* 28 (1990): 399–410, esp. 404; parts of this paper are excerpted in this chapter. See also James F. Veninga, ed., *The Biographer's Gift: Life Histories and Humanism* (College Station, Texas: Texas A & M University Press 1983), 4.
4 Thomas L. Hankins, "In Defence of Biography: The Use of Biography in the History of Science," *History of Science* 17 (1979): 1–16 (on 11).
5 NA: J.A. Macdonald papers, vol. 363. JWD to J.A. Macdonald, 11 December 1879.
6 Richard S. Westfall, *Never At Rest: A Biography of Isaac Newton* (Cambridge: Cambridge University Press 1980).
7 Dean Keith Simonton, *Scientific Genius: A Psychology of Science* (Cambridge: Cambridge University Press 1988), 51–2.
8 "Principal Sir William Dawson, of Montreal," *Leisure Hour* (1886), 614–18 (on 617). Manuscript fragments for Dawson's autobiography (MC: Dawson papers, box 22, folder 18).
9 Dawson, *Fifty Years*, 108–9, 121.
10 JWD to McGill Board of Governors, 23 February 1883.
11 MRB: Pamphlet collection, "Sir William Dawson," memorial service at McGill on 20 November 1899, 7.
12 H.M. Ami, "A Brief Biographical Sketch of Sir John William Dawson," *American Geologist* 26 (1900): 1–57, on 17.
13 Dawson, *Fifty Years*, 290.
14 JWD to Charles Lyell, 27 Oct. 1871.
15 JWD to William C. Williamson, 11 April 1877.
16 Charles E. Moyse to JWD, 4 June 1882.
17 JWD to Charles D. Day, 15 March 1878.
18 JWD to Christopher Dunkin, 23 June 1877.
19 MC: Dawson papers, [Clare Harrington?], "Grandmother," p. 21.
20 Peter Redpath to JWD, 11 April 1883.
21 This is the approach of Charles F. O'Brien, *Sir William Dawson: A Life in Science and Religion* (Philadelphia: American Philosophical Society 1971).
22 MRB: Dawson memorial service.
23 MRB: Pamphlet collection, "Inaugural Discourse of J.W. Dawson," November 1855, 19.

24 MRB: Pamphlet 12, Donald H. MacVicar, "Sir William Dawson: A Character Sketch," quoting a writer in the *Canadian Encyclopedia,* vol. 4: 336.
25 George Mercer Dawson to Anna Dawson Harrington, 14 August 1900. Contemporary biographical sketches include H.M. Ami, "Brief Sketch"; F.D. Adams, "Memoir of Sir J. William Dawson," *Bulletin of the Geological Society of America* 11 (1899): 550–80 (bibliography, 557-80); "John William Dawson," *The Canadian Biographical Dictionary,* 28–32.
26 O'Brien, *Dawson,* 181. Other recent evaluations of Dawson's work include E.A. Collard, "Lyell and Dawson: A Centenary," *Dalhousie Review* 22 (1942): 133–44; William R. Shea's introduction to a reprint of Dawson's *Modern Ideas of Evolution* (New York: Prodist 1977), vii–xxv; and John F. Cornell, "From Creation to Evolution: Dawson and the Idea of Design in the Nineteenth Century," *Journal of the History of Biology* 16 (1983): 137–70.
27 In fact, some of Dawson's scientific work still stands today. See Allen Solem and Ellis L. Yochelson, "North American Palaeozoic Land Snails, with a Summary of Other Palaeozoic Nonmarine Snails," *U.S. Department of the Interior: Geological Survey Professional Paper* no. 1072 (1979): 1–42. F.M. Hueber, *"Psilophyton:* The Genus and the Concept," *International Symposium on the Devonian System* 2 (1967): 815–22.
28 See my "New Directions" for summary of this literature.
29 Leon Edel, *Writing Lives: Principia Biographica* (New York: W.W. Norton & Co. 1984; 1st ed. 1959), 30.
30 See my *Index to the Scientific Correspondence of John William Dawson* (BSHS Monograph 7) (Stanford in the Vale: British Society for the History of Science 1992).
31 Edel, *Writing Lives,* 60 ff.
32 G.M. Dawson to A.D. Harrington, 13 February 1901.
33 G.M. Dawson to A.D. Harrington, 12 February 1901. I am indebted to Brad Lockner for mentioning the important family correspondence on this matter.
34 As Edel suggests, *Writing Lives,* 153.
35 Peter Gay, *Freud for Historians* (New York: Oxford University Press 1985), 150. This is what Bernard C. Meyer calls "reductionism," with its attendant parochial limitations; see "Some Reflections on the Contribution of Psychoanalysis to Biography," *Psychoanalysis and Contemporary Science* 1 (1972): 373–90 (on 374).
36 Miles F. Shore, "Biography in the 1980s," *Journal of Interdisciplinary History* 12 (1981): 89–113 (on 102–5, 98, 109).
37 Thomas A. Kohut, "Psychohistory as History," *American Historical Review* 90 (1986): 336–54 (on 342).
38 Anna Harrington, "Notes for a Biography of JWD," n.d. (MC: Dawson papers, box 64).
39 Clare Harrington, "Grandmother," p.13[a].

40 G.M. Dawson to A.D. Harrington, 15 March 1899.
41 Quoted in Edel, *Writing Lives,* 19.
42 Stanley Frost, *McGill University for the Advancement of Learning* 1: 1801–95 (Montreal: McGill-Queen's University Press 1980); Margaret Gillett, *She Walked Very Warily* (Montreal: McGill-Queen's University Press 1981); and Morris Zaslow, *Reading the Rocks* (Ottawa: Macmillan 1975). Also see Clelia Pighetti, "William Dawson and Scientific Education," *Dalhousie Review* 60 (Winter 1980–81): 622–33.
43 Carl Berger, *Science, God, and Nature in Victorian Canada* (Toronto: University of Toronto Press 1983); A.B. McKillop, *A Disciplined Intelligence: Critical Enquiry and Canadian Thought in the Victorian Era* (Montreal: McGill-Queen's University Press 1979); Suzanne Zeller, *Inventing Canada: Early Victorian Science and the Idea of a Transcontinental Nation* (Toronto: University of Toronto Press 1987).
44 Thomas L. Hankins, "Defence," 2.
45 For example, JWD to Lyell, 27 October 1871.
46 G.M. Dawson to A.D. Harrington, 21 August 1900.
47 Rankine Dawson to G.M. Dawson, 12 January 1900.
48 Dawson, *Fifty Years,* 162.
49 *The Outlook* (London), 25 November 1899.
50 Dawson, *Fifty Years,* 162.
51 Dawson, *Fifty Years,* 162.

CHAPTER 2

1 Portions of this chapter and the next have appeared in my "Sir William Dawson: The Nova Scotia Roots of a Geologist's Worldview," in Paul A. Bogaard, ed., *Profiles of Science and Society in the Maritimes prior to 1914* (Sackville, N.B.: Acadiensis Press 1990), 82–99.
2 John Rutherford, *The Coal-Fields of Nova Scotia* (Newcastle-upon-Tyne: A. Reid 1871), 22.
3 George Patterson, *A History of the County of Pictou, Nova Scotia* (Montreal: Dawson Brothers 1877) [facsimile ed. by MIKA STUDIO (Belleville, Ont.: 1972)], 244–5; Marjory Whitelaw, *Thomas McCulloch: His Life and Times* (Halifax: Nova Scotia Museum 1985), 18. Dollars and pounds were used indiscriminately, with the official exchange rate pegged at $4.86 to the pound. Calculations have been simplified by using $4.00. (Stanley Frost, *McGill University for the Advancement of Learning,* 1: 1801–95 (Montreal: McGill-Queen's University Press 1980), 221, note 1.
4 Charles F. O'Brien, *Sir William Dawson: A Life in Science and Religion* (Philadelphia: American Philosophical Society 1971), 14–15.
5 "Is there a Pictovian type?" *Pictou Advocate,* 20 February 1920 (MC: Dawson papers, box 36, folder 3).

6 W. Stanford Reid, "The Scottish Protestant Tradition," in W. Stanford Reid, ed., *The Scottish Tradition in Canada* (Toronto: McClelland and Stewart 1976), 118–36 (on 132–3). F.C. MacIntosh, "Some Nova Scotia Scientists," *Dalhousie Review* 10 (1930): 199–213 (on 208).
7 James Dawson, "Incidents of a Life" (MC: Papers of James Dawson). For an account of James Dawson's ancestors, his youth in Scotland, and his voyage to Pictou, see Rankine Dawson, ed., *Fifty Years of Work in Canada, Scientific and Educational* (London and Edinburgh: Ballantyne, Hanson & Co. 1901), 5–14.
8 Dawson, *Fifty Years*, 3.
9 For an excellent account of early Pictou see Whitelaw, *McCulloch*, esp. 8–12.
10 James Dawson, "Incidents." See Stanley B. Frost, "A Transatlantic Wooing," *Dalhousie Review* 58 (1978): 458–70, esp. 458–9, for a fine account of James Dawson's life. On James Dawson and the commercial life of Pictou see Patterson, *History*, esp. 305–6, 308.
11 On James Dawson's tribulations see Frost, "Wooing," 458–9. On Pictou's maritime trade see Patterson, *History*, 310–11.
12 Dawson, *Fifty Years*, 14–16.
13 Patterson, *History*, 308.
14 Dawson, Autobiography (MC: Dawson papers, box 22, folder 18).
15 Dawson, Autobiography.
16 Dawson, Autobiography.
17 Dawson, Autobiography.
18 Dawson, Autobiography.
19 Dawson, *Fifty Years*, 26.
20 For a detailed account of the tribulations endured by McCulloch, see the highly subjective account in Patterson, *History*, chap. 16. A more reliable account appears in Whitelaw's exemplary biography of McCulloch.
21 See Patterson, *History*, 324–5. On the unique role of Pictou Academy in early nineteenth-century Nova Scotia, see the list of theses cited in B. Anne Wood, "Thomas McCulloch's Use of Science in Promoting a Liberal Education," *Acadiensis* 17 (1987): 56–73, n. 1.
22 PANS: Pictou Academy papers, "List of students, 1834–35."
23 MacIntosh, "Nova Scotia," 204–5.
24 Offering courses in practical subjects like bookkeeping and navigation was coherent with McCulloch's view that "scientific education" should be grounded in applications to daily life (Wood, "McCulloch's," 63).
25 PANS: Pictou Academy papers, Minutes of trustees' meeting of 7 Aug. 1839.
26 PANS: Pictou Academy papers, Minutes of trustees' meetings of 23 Sept. 1833; 5 Aug. and 30 Nov. 1835.
27 PANS: Pictou Academy papers, Minutes of trustees' meeting of 3 Aug. 1836.

28 Although Anne Wood, "McCulloch's," cites Henry Brougham's publicly expressed "belief" that science was taught at Pictou Academy (56) and "assumes" that McCulloch began to teach science there regularly from 1826 onward (69), she actually provides no evidence to support this view. The tendency to underfund science continued for more than a decade; when the science instructor Charles H. Hay died suddenly in 1849, he was not replaced for a year (Pictou *Advocate*, 21 December 1894 [MC: Dawson papers, box 36, folder 2]). Dawson also recollects in the manuscript version of his autobiography that natural history held "greater charms" for him, but was not formally taught at the academy.

29 "Education in the Pictou Academy," *Eastern Chronicle*, 8 Nov. 1849 (MC: Dawson papers, box 36, folder 2).

30 Dawson extols the importance of the school library in the manuscript version of his autobiography.

31 These and the following remarks are based upon the "Inventory of Books" in the library of the Pictou Academy, taken on 18 May 1845 (PANS: Pictou Academy papers).

32 See Jack Morrell, "The Chemist Breeders: The Research Schools of Liebig and Thomas Thomson," *Ambix* 19 (1972): 1-46 (on 13).

33 Either Robison's *Outlines of a Course of Lectures on Mechanical Philosophy* (1797), his *Elements of Mechanical Philosophy* (1804), or his *Encyclopedia Britannica* articles edited by David Brewster as *A System of Mechanical Philosophy* (1822). The Pictou Library listed both Leslie's *Elements of Natural Philosophy* (1823) and his *Geometry* [*Elements of Geometry, Geometical Analysis, and Plane Trigonometry*] (1809). Two authors who treat these Scottish natural philosophers, among others, are Richard Olson, *Scottish Philosophy and British Physics, 1750-1880* (Princeton: Princeton University Press 1975) and A.L. Donovan, *Philosophical Chemistry in the Scottish Enlightenment: The Doctrines and Discoveries of William Cullen and Joseph Black* (Edinburgh: Edinburgh University Press 1975).

34 DSB: vol. 6: 284.

35 Translations of Gregory's *Astronomiae ... elementa* (1702) and of Keill's *Introductio ad veram physicam* (1701).

36 DSB: vol. 4: 565.

37 DSB: vol. 4: 28.

38 David E. Allen, *The Naturalist in Britain: A Social History* (Harmondsworth: Penguin Books 1978 [1st ed. 1976]), 42.

39 DSB: vol. 12: 588.

40 DSB: vol. 10: 278.

41 R.D. Anderson describes the Scottish interest in a nontechnical logic concerned with method and metaphysics, exemplified in Paley's writings (*Education and Opportunity in Victorian Scotland: Schools and Universities* [Oxford: Clarendon Press 1983], 31).

42 For a description of McCulloch's lectures see Wood, "McCulloch's," 69–72.
43 If this hypothesis is correct, Pictou Academy would have been in the vanguard of the most advanced scientific schools of the continent. There, students had only books to guide their experiments, since practical instruction could be received only in expensive private laboratories. See W.A. Smeaton, "The Early History of Laboratory Instruction in Chemistry at the École Polytechnique, Paris, and Elsewhere," *Annals of Science* 10 (1954): 224. Indeed, Dawson relates in the manuscript version of his autobiography that students were permitted to perform chemistry experiments for amusement in their rooms.
44 See Lewis Pyenson, *Neohumanism and the Persistence of Pure Mathematics in Wilhelmian Germany* (Philadelphia: American Philosophical Society 1983), 11–12. Christa Jungnickel and Russell McCormmach (*Intellectual Mastery of Nature: Theoretical Physics from Ohm to Einstein*, vol. 1 [Chicago: University of Chicago Press 1986]) also argue that little government support went toward the teaching of physical sciences in German universities at this time; professors were even expected to provide their own apparatus to illustrate their lectures (11–23).
45 Whitelaw, *McCulloch*, 32, claims that the academy's laboratory was the only one east of Montreal; I suspect this statement is too modest.
46 The trustees' minutes of 7 June 1832 emphasize that the museum was the property of McCulloch. (PANS: Pictou Academy papers)
47 Whitelaw, *McCulloch*, 35.
48 Dawson, *Fifty Years*, 26, 30.
49 Patterson, *History*, 330. This virtuosity in a variety of fields was part of the Scottish intellectual style; see Anderson, *Education*, 31.
50 Indeed, see the series of letters to the *Eastern Chronicle* (29 Nov. and 20 Dec. 1849), where the young Dawson insists upon freedom of secular education from denominational control.
51 Whitelaw, *McCulloch*, 3.
52 Dawson, *Fifty Years*, 34.
53 Dawson, *Fifty Years*, 36.
54 Dawson, *Fifty Years*, 39–40.
55 Dawson, *Fifty Years*, 41.
56 Frank Dawson Adams, "The History of Geology in Canada," in H.M. Tory, ed., *A History of Science in Canada* (Toronto: Ryerson 1939), 11–12.
57 Bernard J. Harrington, *Life of Sir William E. Logan* (Montreal: Dawson brothers 1883), 175–6, 382–3. Certainly Lyell sounded out Dawson to assist him on an abortive scheme to direct a geological survey of South India (Lyell to JWD, 12 February 1845). He also "exhorted" Logan to employ Dawson as paleontologist to the survey (Lyell to JWD, 21 April 1854).
58 Charles Lyell, *Travels in North America in the Years 1841–2* (New York:

Arno Press, 1978; repr. of 1845 ed.), vol. 1: 3; vol. 2: 136, 173. Lyell to JWD, n.d. [1842?].

59 See Liana Vardi, "J.W. Dawson Correspondence with Sir Charles Lyell, 1842–1875," record group 2d, accession 2081, an unpublished "nearprint" available in the McGill Archives, produced as part of the History of McGill project, iii.

60 Apparently Lyell introduced Dawson to the *Journal* of the Geological Society (EUL: MS gen. 119, JWD to Lyell, 20 Sept. 1845). Also see JWD to Lyell, Nov. 1859.

61 For example, the *QJGS* for 1853 (vol. 10) is launched by two of Dawson's articles (on Nova Scotia coal formations). In 1855 (vol. 11) and 1862 (vol. 18), his articles appear within the first ten pages, where they would be likely to attract the attention of even a casual reader.

62 Lyell to JWD, 2 May 1843, 28 Dec. 1843 [1844?].

63 EUL: JWD to Lyell, Sept. 1845.

64 William Taylor to JWD, 9 June 1842; Lyell to JWD, 17 Nov. 1842.

65 Lyell to JWD, 3 March 1843.

66 Lyell to JWD, 2, 16 May 1843.

67 Courtship letters; see in particular, JWD to Margaret, 18 May, 29 June 1843.

68 JWD to Margaret, 29 Nov. 1841.

CHAPTER 3

1 JWD to Margaret, n.d. [1850?]

2 MC: Dawson papers, [Clare Harrington?], "Grandmother," 1–2, 6.

3 Stanley Frost also draws heavily on the courtship letters in "A Transatlantic Wooing," *Dalhousie Review* 58 (1978): 458–70, although his use of them differs from mine.

4 JWD to Margaret, 5 Aug. 1844.

5 JWD to Margaret, 15 Sept. 1841; 25 Dec. 1845; 27 March, 1 Oct. 1846.

6 JWD to Margaret, 15 June 1844.

7 JWD to Margaret, 10 Sept. 1844.

8 JWD to Margaret, 22 Jan. 1847.

9 JWD to Margaret, 25 Dec. 1845.

10 JWD to Margaret, 7 Jan. 1844.

11 Clare Harrington, "Grandmother," 8.

12 Clare Harrington, "Grandmother," 9.

13 Clare Harrington, "Grandmother," 10.

14 JWD to Margaret, 8 July 1850.

15 MC: Dawson papers, box 1.

16 JWD to Margaret, 19 Jan.; 9 and 13 Feb. 1850.

17 MC: Dawson papers, box 37: Alexander Forrester *et. al.* to JWD, 29 March 1850; Halifax students to JWD, 24 April 1850.

18 JWD to Margaret, 30 Jan. 1850. As the city is described by Marjory Whitelaw, *Thomas McCulloch: His Life and Times* (Halifax: Nova Scotia Museum 1985), 33.
19 PANS: MG 1, vol. 1469, folder 13: JWD to Joseph Howe, 17 April 1850.
20 JWD to Margaret, 23 Feb. 1850; n.d.[1850].
21 PANS: MG 1, vol. 439, folder 60: James Dawson to Howe, 25 May 1850.
22 One of Howe's strong selling-points was that Dawson's travels through Nova Scotia as superintendent would allow him to continue his geological researches. See E.A. Collard, "Lyell and Dawson: A Centenary," *Dalhousie Review* 22 (1942): 133–44.
23 PANS: MG 1, vol. 1469, folder 14: JWD to Howe, 2 May 1850.
24 JWD to Margaret, n.d. [Spring?], 1844.
25 Dawson, Autobiography (MC: Dawson papers, box 22, folder 18). Many years later, Margaret Dawson recollected that James Cosmo had died from a bean lodged in his lung following a choking incident (Clare Harrington, "Grandmother"). However, six months after the baby's death, William told Margaret that the smallpox epidemic was just then abating in Halifax; her concern that he be vaccinated against the disease suggests that smallpox might have carried off their child (JWD to Margaret, 19 Jan., 23 Feb. 1850). On at least one occasion, Margaret left her child in the care of her in-laws and accompanied William on his tour of duty.
26 JWD to Margaret, 10 July 1850.
27 Rankine Dawson, ed., *Fifty Years of Work in Canada, Scientific and Educational* (London and Edinburgh: Ballantyne, Hanson & Co. 1901), 84–5.
28 PANS: MG 1, vol. 1469, folder 33.
29 PANS: MG 1, vol. 1469, folder 28, James Dawson to Howe, 8 Dec. 1851.
30 PANS: MG 1, vol. 1469, folder 29, JWD to Howe, n.d. [1852].
31 John P. Vaillancourt, "John William Dawson, Education Missionary in Nova Scotia" (M.Ed. thesis, Dalhousie University 1973), 1.
32 Dawson, *Fifty Years*, 79.
33 JWD to Margaret, 14 Sept. 1850.
34 Lyell to JWD, 25 March 1853. Owen would have appended *Lyellii*, but Lyell preferred to commemorate Nova Scotia in the nomenclature.
35 Lyell returned to Nova Scotia at the end of July 1852; he wanted to spend a few days exchanging ideas and geologizing with Dawson (MC: Lyell to JWD, 9 July 1852). See Katherine M. Lyell, ed., *Life, Letters and Journals of Sir Charles Lyell Bart* (London: John Murray 1881), vol. 2: 181–2, on paleobotanical investigations with Dawson in 1852; on the discovery of the coal reptile, 183, 185–6. Lyell crossed the Atlantic with Edmund Head and his wife, whom he introduced to Dawson.
36 Lyell to JWD, 12 Nov. 1852.
37 Lyell to JWD, 15 July 1852.

38 "On the Structure of the Albion Mines Coal Measures, Nova Scotia," *QJGS* 10 (1854): 42–47; "On the Coal-Measures of the South Joggins, Nova Scotia," *QJGS* 10 (1854): 1–42. Lyell to JWD, 17 Dec. 1852, 17 May 1853.
39 John Ross to JWD, 1 May 1846.
40 J.S. Martell, "Early Coal Mining in Nova Scotia," *Dalhousie Review* 25 (1945–46): 156–72 (on 159).
41 Richard Brown, *The Coal Fields and Coal Trade on the Island of Cape Breton* (London: Sampson Low 1871), 101.
42 I am indebted to Donald Macleod's "Faith, Hope and Geology: Practical Geology in Nineteenth Century Nova Scotia," delivered to the Science and Society in the Maritimes conference, Mount Allison University, Sept. 1988, for a general understanding of Nova Scotia mining; see esp. p. 8.
43 JWD to G.M. Young, 20 April 1848.
44 JWD to G.M. Young, 27 Feb. 1845. S. Cunard to JWD, 28 Nov. 1845; Richard Brown to JWD, 9 Dec. 1845.
45 R.W. Ells, "The Progress of Geological Investigation in Nova Scotia," *Proceedings of the Nova Scotia Institute of Natural Science* 10 (1898–1902): 433–46 (on 440).
46 Macleod, "Faith," 4.
47 Macleod, "Faith," 10.
48 Macleod, "Faith," 10.
49 G.M. Young to JWD, 1 March 1848.
50 Bernard C. Meyer cites many instances where creative work functions as a "grateful monument" to the memory of a loved one; see "Some Reflections on the Contribution of Psychoanalysis to Biography," *Psychoanalysis and Contemporary Science* 1 (1972): 373–90 (on 384).
51 Dawson's father had become proprietor of a printing establishment in 1835 (Dawson, *Fifty Years*, 31).
52 JWD to Oliver & Boyd, 26 Feb. 1855.
53 Lyell to JWD, 9 July, 12 Aug. 1852.
54 On Lizars, see my "From the North to Red Lion Court: the Creation and Early Years of the *Annals of Natural History*," *Archives of Natural History* 10 (1981): 221–49 and my "War and Peace in Natural History Publishing: The *Naturalist's Library*, 1833–1843," *Isis* 72 (1981): 50–72.
55 Presumably his *Guide to the Geology of Scotland* of 1844.
56 Oliver and Boyd to JWD, 25 Aug. 1853, 16 March 1855; JWD to Oliver & Boyd, 26 Feb., 10 April 1855.
57 Invoice from Lizars to JWD, 30 Dec. 1856; Oliver & Boyd to JWD, 6 July 1855. R. Brown to JWD, 4 Jan. 1855; John S. Saunders to JWD, 6 July 1855.
58 F.W. Gray, "Pioneer Geologists of Nova Scotia," *Collections of Nova Scotia Historical Society* 26 (1945): 153–72 (on 162).
59 Dawson, *Acadian Geology: An Account of the Geological Structure and Mineral Resources of Nova Scotia, and Portions of the Neighbouring Provinces of*

British America (Edinburgh: Oliver and Boyd; London: Simpkin, Marshall, and Co.; Pictou, N.S.: J. Dawson and Son 1855), 1–2.
60 *Acadian Geology*, 2–3.
61 *Acadian Geology*, 3–4, 6–11.
62 *Acadian Geology*, 365.
63 *Westminster Review* 64 (1855): 569–73.
64 *Athenaeum*, no. 1470 (29 Dec. 1855): 1527–8.
65 Lyell to JWD, n.d. [1855?].
66 Philip Egerton to JWD, 19 Sept. 1855.
67 Hew Ramsay to James Dawson, 21 Aug. 1855.
68 JWD to Margaret, 10 Sept. 1844.
69 MC: Dawson papers, James Dawson's "Narrative Diary, 1849–61."
70 Whitelaw, *McCulloch*, 36.

CHAPTER 4

1 I have published this chapter in a slightly different form as "Horse Race: John William Dawson, Charles Lyell, and the Competition over the Edinburgh Natural History Chair in 1854–1855," *Annals of Science* 49 (1992): 461–77.
2 I have borrowed heavily from the imagery of horseracing, because this seems to be an appropriate analogy for describing the competition over the Edinburgh chair (indeed, one contemporary journalist used the metaphor in the *Scottish Press*). Especially enlightening is Marvin B. Scott's *The Racing Game* (Chicago: Aldine Publishing Co. 1968), which relates the characteristics of the race course to sociological institutions.
3 Recounted in Laurence Hutton, *Literary Landmarks of the Scottish Universities* (New York: G.P. Putnam's Sons 1904), 68. (See Nora Barlow, ed., *The Autobiography of Charles Darwin* [London: Collins 1958], 52–3). Other bad memories of Jameson are mentioned in James Secord, "The Discovery of a Vocation: Darwin's Early Geology," *British Journal for the History of Science* 24 (1991): 133–57, although he argues that these criticisms are misleading. Secord maintains that Jameson's course was the most comprehensive and best attended in Britain (134–5).
4 J.B. Morrell, "Science and Scottish University Reform: Edinburgh in 1826," *British Journal for the History of Science* 6 (1972): 39–56 (on 48, 50).
5 George Wilson and Archibald Geikie, *Memoir of Edward Forbes, F.R.S.* (Cambridge and London: Macmillan 1861), 109. The laudatory obituary published by Jameson's nephew, Laurence, lists a number of his eminent pupils; see "Biographical Memoir of the Late Professor Jameson," *Edinburgh New Philosophical Journal* 62 (1854): 1–49 (on 3, 31).
6 Morrell, "Science," 40; Secord, "Discovery," 135.
7 Both James Scotland, *The History of Scottish Education* (London: University

of London 1969), vol. 1: 351–2, and D.B. Horn, *A Short History of the University of Edinburgh* (Edinburgh: Edinburgh University Press 1967), 96, state that no new chairs were created in arts from 1760 to 1862. Astronomy and technology professorships, however, were founded during this period in the arts faculty.

8 R.D. Anderson, *Education and Opportunity in Victorian Scotland: Schools and Universities* (Oxford: Clarendon Press 1983), 34.

9 John Campbell Shairp, Peter Guthrie Tait, and A. Adams-Reilly, *Life and Letters of James David Forbes* (London: Macmillan 1873), 130–1.

10 George Elder Davie, *The Democratic Intellect: Scotland and her Universities in the Nineteenth Century* (Edinburgh: Edinburgh University Press 1961), 106.

11 Davie, 190. J.B. Morrell has also described how metaphysical issues served as a smokescreen to obscure political and ecclesiastical differences during the succession to the mathematics chair early in the century; see his "The Leslie Affair: Careers, Kirk and Politics in Edinburgh in 1805," *Scottish Historical Review* 54 (1975): 62–82.

12 John Kerr, *Scottish Education: School and University from Early Times to 1908* (Cambridge: Cambridge University Press 1910), 265–7.

13 Anderson, *Education*, 28–9. J.B. Morrell shows that for the period 1871–76, natural history was second only to anatomy and chemistry in terms of total emoluments (salary plus student fees) brought in by the chair ("The Patronage of Mid-Victorian Science in the University of Edinburgh," *Science Studies* 3 [1973]: 353–88, table 5).

14 According to Morrell, four of the fifteen chairs in science and medicine were Crown appointments; of these, only the astronomy chair (founded 1786) resided in the Faculty of Arts. The Faculty of Medicine, then, could claim three Regius professorships: natural history (founded 1767), clinical surgery (founded 1803), and medicial jurisprudence (founded 1807) (Morrell, "Patronage," table 2).

15 OUM: James D. Forbes to Phillips, 20 Nov. 1854. (Forbes's endurance seems somewhat less remarkable when contrasted with the career of John Leslie, who tried for five Scottish university chairs; see Morrell, "Leslie Affair," 67.)

16 In a letter submitting his candidacy for Jameson's chair (anticipating his retirement), the zoologist John Fleming, then aged 68, claimed that his only opponent "talked of" in Edinburgh was many years older (SRO: AD 56/49, Fleming to the Lord Advocate, 10 Oct. 1853). Apparently Fleming was alluding to T.S. Traill (see GS: James Nicol to Roderick Murchison, 5 May 1854). Fleming had held the chair in natural philosophy at Aberdeen; he left it for the natural science chair at New College, Edinburgh.

17 It was highly unusual for the Town Council to act in concert with the university, as it did in the case of Forbes. See SRO: "Memorial for the

Lord Provost, Magistrates and Council of the City of Edinburgh to Viscount Palmerston," 24 April 1854.
18 Wilson and Geikie, *Memoir,* 307, 526, 552.
19 EUL: Senate Minutes, 15 May 1854.
20 Wilson and Geikie, *Memoir,* 570.
21 Scott (*Racing,* 9) describes the paddock (a possible feature of all organizations) as a place where intentions may be gleaned.
22 Wilson and Geikie, *Memoir,* 555.
23 Lyell to JWD, 18 Dec. 1854.
24 Lyell to JWD, 29 Dec. 1854; also see Wilson and Geikie, *Memoir,* 327. On Fleming, see n.16.
25 "The Natural History Chair," *Scottish Press,* July 1855.
26 OUM: Forbes to Phillips, 20 Nov. 1854.
27 Lyell to JWD, 18 Dec. 1854; Jules Marcou, *Life, Letters and Works of Louis Agassiz* (New York: Macmillan 1896), vol. 2: 69, 71–2. Edward Lurie mentions the offer from Edinburgh to Agassiz, which he claims did not interest Agassiz; however, no trace of the offer appears in the life and letters (an earlier collection was edited by E. Agassiz) or in the Agassiz papers at the Museum of Comparative Zoology. See *Louis Agassiz: A Life in Science* (Baltimore: Johns Hopkins Press 1988; 1st ed. 1960), 192. In the Edinburgh Town Council minutes, a motion to petition the Home Department in Agassiz's favour was rescinded after much discussion (ETC: vol. 265, 27 Feb. [47–8] and 13 March 1855 [98]). According to a letter from one of the councillors, John Robertson, to the lord advocate, he withdrew the motion because he feared it would have been defeated and the rebuff communicated to Agassiz (SRO: 13 March 1855).
28 Wilson and Geikie, *Memoir,* 555.
29 OUM: Forbes to Phillips, 20 Nov. 1854.
30 Leonard Huxley, ed., *Life and Letters of Thomas Henry Huxley* (London: Macmillan 1903), vol. 1: 170, 173; ICS: Huxley to J.D. Hooker, 24 Nov. 1854.
31 Wilson and Geikie, *Memoir,* 562, 565; Lyell to JWD, 18 Dec. 1854.
32 *Scottish Press,* "Chair."
33 Wilson and Geikie, *Memoir,* 404, 523. James Nicol pronounced Laurence Jameson "dull" in a letter to Roderick Murchison (GS: 5 May 1854).
34 Wilson and Geikie, *Memoir,* 559; Lyell to JWD, 18 Dec. 1854. No direct evidence exists that Hugh Miller was a candidate in this run, although he had earlier hoped for the chair that went to Forbes. Nevertheless, he was at the height of his fame as a scriptural geologist at this time, his autobiographical *My Schools and Schoolmasters* having just been published. Peter Bayne, ed., *Life and Letters of Hugh Miller* (London: Strahan and Co. 1871), vol. 2: 436, 445–6; George Rosie, *Hugh Miller: Outrage and Order: A Biography and Selected Writings* (Edinburgh: Mainstream Publishing Co.

n.d. [1981?], 9, 76). After Forbes's death, Huxley was willing to stand for the chair, but felt his chances were slim and that the professorship would be divided (ICS: T.H. Huxley to J.D. Hooker, 24 Nov. 1854). The Town Council moved to divide the chair on 5 Dec. but rejected the motion on 19 Dec. 1854 (ETC: vol. 264, 197, 228).

35 L.S. Jacyna, ed., *A Tale of Three Cities: The Correspondence of William Sharpey and Allen Thomson (Medical History Supplement*, no. 9 [1989]), xxi.

36 n.a., *Edinburgh University: a Sketch of its Life for 300 Years* (Edinburgh: James Gemmell 1884), 76–8.

37 *Edinburgh University: A Sketch*, 78.

38 Horn, *Short History*, 152.

39 Lyell to JWD, 20 Jan. 1855.

40 Lyell to JWD, 18 Dec. 1854, 20 Jan. 1855.

41 Lyell to JWD, 27 Nov., 29 Dec. 1854.

42 SRO: Lyell to [Lord Advocate], 1 Oct. 1853.

43 Lyell to JWD, 27 Nov. 1854.

44 Of course, this was the only means of communicating urgent information. In winter, Dawson could not respond by return mail, because he did not receive mail in Pictou until after the departure of the return boat. At one point, he resorted to asking William Young, the provincial attorney general, to open his mail in Halifax (JWD to Lyell, 2 and 16 Jan. 1855).

45 JWD to Lyell, 2 Jan. 1855.

46 JWD to Lyell, 16 Jan. 1855.

47 Lyell to JWD, 27 Nov. 1854.

48 Lyell to JWD, 18 Dec. 1854.

49 Alexander Forrester to JWD, 20 Jan. 1855.

50 JWD to Lyell, 2 Jan. 1855.

51 JWD to William Chambers, 16 Jan. 1855; JWD to Lyell, 2 Jan. 1855; Lyell to JWD, 2 Feb. 1855.

52 JWD to Lyell, 2 Jan. 1855.

53 See Johnston's formal petition, entitled "A Letter to Viscount Palmerston," concerning "The Chair of Natural History in the University of Edinburgh," and his discussion in *Selections from the Correspondence of Dr. George Johnston*, ed. by James Hardy (Edinburgh: David Douglas 1892), 520 ff.

54 Lyell to JWD, 20 Jan. 1855.

55 See Huxley, *Life and Letters*, 174; apparently Huxley was approached by several Edinburgh professors in Nov. and Dec. 1854, and again in the following spring (170, 173–4). Lyell to JWD, 2 Feb. 1855. An obituary notice of George James Allman (1812–1898) appears in the *Proceedings of the Royal Society* 75 (1905): 25–7. Already in February, he had asked the eminent French zoologist Henri Milne-Edwards for a recommendation (EUL: Allman to Milne-Edwards, 8 Feb. 1855).

56 Alexander Forrester to JWD, 20 Jan. 1855.
57 Lyell to JWD, 2 Feb. 1855.
58 Indeed, one historian views the lecture format as a kind of "lay preaching," which worked because of the Scottish people's familiarity with sermons. He adds that students were "inured to tedium and appreciative of the finer points of oral exposition" (Anderson, *Education*, 32).
59 Peter Bell to JWD, 27 April 1855.
60 SRO: Lyell to Lord Advocate, 29 June 1855.
61 Lyell to JWD, 18 Dec. 1854.
62 Lyell to JWD, 2 and 24 Feb. 1855.
63 Lyell to JWD, 7 Aug. 1855.
64 Lyell to Henry De La Beche, 17 March 1855.
65 Lyell to JWD, 24 Feb. 1855.
66 Lyell to JWD, 29 Dec. 1854.
67 GS: James Nicol to Murchison, 27 Nov. 1854.
68 SRO: Allman to Lord Advocate, 6 March 1855; *Testimonials in favour of George James Allman*.
69 Bell to JWD, 16 Feb. 1855.
70 Lyell to JWD, 24 Feb. 1855.
71 Lyell to JWD, 20 April 1855.
72 Lyell to JWD, 20 April 1855; JWD to Lyell, 5 June 1855.
73 JWD to Lyell, 13 March 1855.
74 Bell to JWD, 13 April 1855.
75 ETC: vol. 265, 24 April 1855, 233–4.
76 Bell to JWD, 27 April 1855.
77 Lyell to JWD, 25 May 1855.
78 ICS: Allman to Huxley, 21 April, 20 May 1855. Even the Town Council registered its displeasure over the delay in filling the chair, and readied a delegation to present a memorial to the home secretary in London (ETC: vol. 265, 3 July 1855, 416; vol. 266, 17 July 1855, 16–20).
79 Lyell to JWD, 2 July 1855.
80 Cited in Bell to JWD, n.d. [July 1855?].
81 SRO: Lyell to Lord Advocate, 2 June 1855.
82 SRO: Lyell to Lord Advocate, 29 June 1855.
83 SRO: Leonard Horner to W.P. Alison, 2 July 1855.
84 SRO: Horner to Alison, 2 July 1855.
85 SRO: Alison to Horner, 5 July 1855.
86 Bell to JWD, 5 July 1855 and n.d. [July 1855?].
87 Bell to JWD, 20 July 1855. Apparently the decision was announced by Whitehall on 10 July 1855; see ETC: vol. 266, 20–1.
88 Lyell to JWD, 19 Aug. 1855. EUL: Senate Minutes of 5 Nov. 1855.
89 *Scottish Press*, "Chair."
90 "The Natural History Chair," *Scotsman*, 14 July 1855.

230   Notes to pages 52–6

91   Lyell to JWD, 19 Aug. 1855.
92   ICS: Lyell to Huxley, 29 Aug. 1855.
93   JWD to Margaret, 14 Sept. 1855.
94   Bell to JWD, 20 Dec. 1855.
95   Oliver & Boyd to JWD, 2 March 1855.
96   Dawson, Autobiography (MC: Dawson papers, box 22, folder 18).

CHAPTER 5

1   JWD to Margaret, n.d. [summer] 1850.
2   MC: Dawson papers, box 22, folder 18, Autobiography.
3   From the beginning, McGill College constituted a university; the two terms are therefore used interchangeably here. After 1885, "McGill College" dropped out of general usage, resulting in the title, "McGill University." (Stanley Frost, *McGill University for the Advancement of Learning*, 1: 1801–95 (Montreal: McGill-Queen's University Press 1980), 49–50.
4   Rankine Dawson, ed., *Fifty Years of Work in Canada, Scientific and Educational* (London and Edinburgh: Ballantyne, Hanson & Co. 1901), 90–2. (Dawson was awarded an M.A. by the University of Edinburgh in 1856.)
5   Joseph Howe to JWD, 18 May 1855.
6   Dawson, Autobiography.
7   Donald MacKay, *The Square Mile: Merchant Princes of Montreal* (Vancouver: Douglas & McIntyre 1987), 43.
8   MRB: Dawson, "Inaugural Discourse," 3.
9   Dawson, Autobiography, and *Fifty Years*, 91–2.
10 Dawson, *Fifty Years*, 98–99.
11 Dawson, Autobiography.
12 ICS: Lyell to T.H. Huxley, 29 Aug. 1855.
13 MC: Dawson papers, [Clare Harrington?], "Grandmother," 12–13.
14 JWD to James Dawson, 3, 10, 17, 24 Nov.; 8 Dec. 1855; 19 Jan., Feb. 1856.
15 Frost, *McGill*, explores Dawson's early career as an educationalist, 181–4. He also describes how McGill's curriculum was remodeled in 1852 after those of the Scottish universities, thereby definitively rejecting the Oxbridge legacy. (See also his *History of McGill in Relation to Montreal and Quebec* [Montreal: McGill University 1979], 4).
16 MRB: Pamphlet collection, "Sir William Dawson," memorial service at McGill, 7.
17 As the method was later explained in George Kennedy to JWD, 27 July 1880.
18 SIA: JWD to Baird, 9 Feb. 1856.
19 JWD to James Dawson, 24 Nov., 8 Dec. 1855. C.D. Day to JWD, 20 Aug. 1855.
20 JWD to James Dawson, 16 Jan. 1855.

21 MC: Dawson papers, Clare Harrington, "Grandfather," 7.
22 Dawson, Autobiography.
23 Frost, *McGill*, 199–200.
24 Frost, *History*, 7.
25 See MacKay, *Square Mile*, and Margaret Westley, *Remembrance of Grandeur: the Anglo-Protestant Elite of Montreal, 1900–1950* (Montreal: Libre Expression 1990), 7.
26 See Robert Falconer, "Scottish Influence in the Higher Education of Canada," *Trans. Roy. Soc. Can.* Sect 2 (1927): 7–20 (on 10 ff.)
27 Daniel Wilson to JWD, 21 March 1871.
28 Dawson, Autobiography, and *Fifty Years*, 111.
29 H.H. Wood to JWD, 3 Jan. 1882.
30 R.A. Ramsay to JWD, 18 Feb. 1882.
31 Dawson, *Fifty Years*, 105; Frost, *McGill*, 200.
32 Peter Redpath to JWD, 13 July 1881.
33 Redpath to JWD, 9 Nov. 1881.
34 Redpath to JWD, 4 April 1884.
35 For example, see Redpath to JWD, 12 April 1885.
36 JWD to Edward Greenshields, 18 March 1886.
37 William Workman to JWD, 10 Feb. 1870.
38 JWD to Charles Day, 9 March 1876. Abbott was appointed governor in 1881.
39 See Frost, *McGill*, 201–2, 240.
40 R.A. Ramsay to JWD, 21 Dec. 1882; C.D. Day to JWD, 4 Jan., 5 March 1883.
41 Alexander Johnson to JWD, 28 May 1870; Dawson, *Fifty Years*, 163.
42 Dawson, *Fifty Years*, 162–3.
43 Frost, *McGill*, 210.
44 JWD to Lyell, 27 Dec. 1860.
45 Frost, *McGill*, 213.
46 G.J. Jarish to JWD, 17 Sept. 1869.
47 JWD to Henry Bovey, 11 Dec. 1876.
48 Clipping from E.J. Collard's "All our Yesterdays" in MC: Dawson papers, box 37.
49 Mary Dunkin to JWD, 6 April 1870.
50 C.D. Day to JWD, 20 Aug. 1855.
51 MRB: Dawson, "Inaugural discourse," 18.
52 JWD to James Ferrier, 4 Oct. 1883.
53 See Suzanne Zeller, *Inventing Canada: Early Victorian Science and the Idea of a Transcontinental Nation* (Toronto: University of Toronto Press 1987), who explores how these three disciplines shaped the Canadian identity.
54 Dawson, Autobiography.
55 Frost, *McGill*, 222–3.
56 Frost, *McGill*, 186, 269. Richard A. Jarrell, *The Cold Light of Dawn: A*

History of Canadian Astronomy (Toronto: University of Toronto Press 1988), 46. I discuss the Natural History Society of Montreal in chap. 12.
57 Malcolm M. Thomson, *The Beginning of the Long Dash* (Toronto: University of Toronto Press 1978), 7–8.
58 Jarrell, *Cold Light*, 47–8.
59 G.T. Kingston to JWD, 30 Dec., 31 Dec. 1873.
60 Frost, *McGill*, 269–70. Jarrell, *Cold Light*, 48. JWD to Andrew Robertson, 10 Sept. 1879. Thomson, *Beginning*, 8–9.
61 JWD to Minister of Marine and Fisheries, Jan. 1882. Jarrell, *Cold Light*, 50.
62 Thomson, *Beginning*, 30; Jarrell, *Cold Light*, 68.
63 Frost, *McGill*, 270, 273.
64 JWD to Charles Day, 1 March 1870.
65 Frost, *McGill*, 186. Dawson published a textbook on agriculture: *First Lessons in Scientific Agriculture for Schools and Private Instruction* (Montreal: 1864), based on his earlier *Scientific Contributions towards the Improvement of Agriculture* (Pictou: 1853).
66 JWD to Joseph Perrault, 18 March 1864. See Ralph H. Estey, *Essays on the Early History of Plant Pathology and Mycology in Canada* (Montreal and Kingston: McGill-Queen's University Press 1994).
67 Frost, *McGill*, 188.
68 JWD to C.D. Day, 15 April 1869. For example: John Rose to JWD, 2 June 1869; Joseph Howe to JWD, 16 Dec. 1869; E.A. Meredith to JWD, 1 Feb. 1870. JWD to P.J. Chauveau, 28 May 1871.
69 JWD to Rose, 13 April 1869.
70 Christopher Dunkin to JWD, 6 Jan. 1870.
71 Frost, *McGill*, 188, 214.
72 JWD to Hugh Allan, Aug. 1870.
73 Bernard Harrington to JWD, 26 Feb. 1871.
74 George Mercer Dawson to JWD, 25 June 1871.
75 Frost, *McGill*, 274.
76 JWD to J.D.S. Campbell, 19 March 1880. For example: Hugh McLennan to JWD, 7 March 1878; Peter Redpath to JWD, 9 April 1878. I discuss the survey's move to Ottawa in chap. 6.
77 B.J. Harrington to JWD, 19 July 1883.
78 *Hand-book for the City of Montreal and its Environs, prepared for the meeting of the AAAS, at Montreal* (Montreal: Gazette Printing Co. 1882)
79 Frost, *McGill*, 274, 277.
80 Thomas Cramp to JWD, 10 Feb. 1870. JWD to William Logan, 28 April and 7 July 1870.
81 Logan to JWD, 26 April 1870; Charles Day to JWD, 28 Feb. 1872.
82 William C. Baynes to JWD, 10 July 1883.
83 JWD to Charles Gibb, 10 April 1883.
84 Gibb to JWD, 13 April 1883.

85 Asa Gray to JWD, 31 March 1883.
86 John Macoun to JWD, 4 April 1883.
87 George F. Matthew to JWD, 5 and 14 April 1883; L.W. Bailey to JWD, 9 April 1883; James Vroom to JWD, 9 April 1883.
88 Bernard Harrington to JWD, 21 Sept. 1883; Anna Dawson to JWD, 2 Nov. 1883; 14 Jan. 1884.
89 David Penhallow to JWD, 4 Dec. 1883.
90 JWD to Spencer Baird, 3 Nov. 1855.
91 For much of this account, see my "'Stones and Bones and Skeletons': The Origins and Early Development of the Peter Redpath Museum," *McGill Journal of Education* 17 (1982): 45–64, and my *Cathedrals of Science: The Development of Colonial Natural History Museums during the Late Nineteenth Century* (Montreal: McGill Queen's University Press 1988).
92 JWD to William Logan, 9 Dec. 1867.
93 Dawson, *Fifty Years*, 169–72. SIA: JWD to Joseph Henry, 5 Jan. 1869. JWD to W.T. Thiselton Dyer, 7 July 1877.
94 JWD to Anne Molson, 18 Nov. 1872.
95 Dawson, *Fifty Years*, 173.
96 "Notice of the Natural History Collections of the McGill University," *CN* 7 (1862): 221–3. SIA: JWD to Joseph Henry, 5 Jan. 1869.
97 "Address to the Seventh Annual Conversazione," *CN* ns 4 (1869): 67.
98 I discuss the move of the museum in chap. 6.
99 *Witness*, 28 Feb. 1881. JWD to Charles Tupper, 12 Feb. 1881; "The Geological Survey," *Gazette* [?], 1881.
100 See chap. 6 on this event.
101 Peter Redpath to JWD, 19 Jan. 1880; Charles Day to JWD, 16 Feb. 1880. Frost, *McGill*, 243; J.W. Dawson, *In Memoriam: Peter Redpath, Governor and Benefactor of McGill* (Montreal: "Witness" Printing House 1894), 15. Redpath also provided $10,000 for the purchase of display cases.
102 B.J. Harrington to JWD, 19 Aug. 1880.
103 JWD to Henry Ward, 22 Jan. 1881.
104 Dawson, *Fifty Years*, 176. G.R. Grant to JWD, 10 June 1882.
105 Molson's wife Louisa supplied the salary of Thomas Curry, the assistant curator. Dawson, *Fifty Years*, 176.
106 G.M. Dawson to JWD, 20 and 26 June, 17 and 23 July, 24 Sept. 1882. JWD to James Ferrier, 3 April 1884.
107 Robert Hamilton to JWD, 12 Aug. 1881; A.H. Foord to JWD, 13 Nov. 1882; JWD to Charles Tupper, 12 Feb. 1881.
108 JWD to T.H.S. White, 17 Feb. 1881. Spencer Baird to JWD, 10 and 17 May, 4 June 1883.
109 Peter Redpath to JWD, 17 Jan. 1881. *Report on the Peter Redpath Museum* 2 (Jan. 1883): 1–2. P. Kuetzing to JWD, 12 Sept. 1882. Sally Kohlstedt, "Henry A. Ward: The Merchant Naturalist and American Museum

Development," *Journal of the Society for the Bibliography of Natural History* 9 (1980): 647–61 (on 650).
110 Dawson, *Fifty Years*, 174; "The American Association: Opening the Redpath Museum," *Witness*, 25 Aug. 1882; "The Peter Redpath Museum: Formal Opening of the Building," clipping dated 25 Aug. 1882 in MC: Dawson papers, box 37.
111 Dawson, *Fifty Years*, 174. Dawson, *In Memoriam: Redpath*, 15, 17.
112 *Guide to Visitors to the Peter Redpath Museum of McGill University* (Montreal: 1885), 2.
113 HUM: JWD to Alexander Agassiz, 18 July 1882.
114 This and subsequent information on the administration, finances, and management of the Peter Redpath Museum comes from the museum's Minute Book for 1882–92 (MC: Acc. 1602, 1b).
115 After reducing the fee to 10 cents, the Minute Book claimed, however, that it was "not imposed for revenue" (128).
116 Minute Book, 44, 48–9, 95.
117 Minute Book, 30, 115, 51, 66.
118 Minute Book, 81, 92, 103–4.
119 MC: Corporation Minutes, 189–94, p. 392. Minute Book, 107.
120 Peter Redpath to JWD, 12 April 1885. For information on the Peter Redpath Museum subsequent to Dawson's retirement as principal, see my "'Stones and Bones and Skeletons'" and my *Cathedrals of Science*.
121 *Times*, 20 Nov. 1899.
122 C.D. Day to JWD, 23 Oct. 1883.
123 J.M. Lemoine to JWD, 31 March 1883.
124 Edmund Head to JWD, 27 Oct. 1861.

CHAPTER 6

1 MC: Logan papers, JWD to Logan, 6 Jan. 1865.
2 J.W. Dawson, *On Some Points in the History and Prospects of Protestant Education in Lower Canada* (Montreal: J.C. Becket 1864), 19–20.
3 Rankine Dawson, ed., *Fifty Years of Work in Canada, Scientific and Educational* (London and Edinburgh: Ballantyne, Hanson & Co. 1901), 164.
4 Dawson, *Fifty Years*, 115–18, 121. Also see Stanley Frost, *McGill University for the Advancement of Learning*, 1: 1801–95 (Montreal: McGill-Queen's University Press 1980), 188–93.
5 Dawson, *Fifty Years*, 122–3.
6 See Harry Kuntz and Calvin Potter, "Whither the Protestant School System in Quebec?," [QFHSA's] *Home and School News* (Dec. 1989): 7–12, for a good condensed summary of the history of education in Quebec (esp. 7–8). Also see their *Whither the Protestant School System in Quebec?* (Montreal: QFHSA 1989).

7 See Frost, *McGill*, 195–6.
8 From 1867–74, a "minister" replaced the "superintendent."
9 Kuntz and Potter, "Whither," 8.
10 MC: Logan papers, JWD to Logan, 28 April 1865; 15 March 1867.
11 MC: Logan papers, JWD to Logan, 6 Jan. 1865.
12 JWD to Lyell, 12 Dec. 1866.
13 JWD to Lyell, 12 Dec. 1866; MC: Logan papers, JWD to Logan, 6 Jan. 1865.
14 JWD to Lyell, 16 March 1865.
15 William James Anderson to JWD, 13 Dec. 1869.
16 Dawson, *Protestant Education*, 9–12.
17 H.H. Miles to JWD, 26 Feb. 1870.
18 De Sola to JWD, 9 June 1870.
19 Adam Shortt and Arthur G. Doughty, eds., *Canada and its Provinces*, (Toronto: Glasgow, Brook & Co. 1914), vol. 16, pt. 2: 480.
20 P.J. Jolicoeur to JWD, 3 March 1876. See also Louis-Philippe Audet, *Histoire du Conseil de l'instruction publique de la province de Québec 1856–1964* (Montréal: Éditions Leméac 1964), 97. Also sitting with Dawson were James W. Williams (the Anglican Bishop of Quebec), Charles D. Day, Christopher Dunkin, George Irvine, Reverend John Cook, Archdeacon William T. Leach, and James Ferrier.
21 Shortt and Doughty, *Canada*, 475.
22 C.D. Day to JWD, 24 June 1876.
23 James Ferrier to JWD, 4 Dec. 1875; F.C. Emberson to JWD, 15 Jan. 1876.
24 H.H. Miles to JWD, 18 April 1876.
25 JWD to George B. Baker, 14 July 1876; Baker to JWD, 4 July 1876.
26 Audet, *Histoire*, 176.
27 George Weir to JWD, 3 March 1877.
28 C.D. Day to JWD, 16 Jan. 1880.
29 JWD to Lyell, 10 Oct. 1868.
30 R.W. Norman to JWD, 30 Aug. 1883.
31 R.W. Heneker to JWD, 31 Oct. 1885.
32 J.A. Macdonald to JWD, 15 June 1889.
33 T.H. Rand to JWD, 7 April 1869.
34 MC: Logan papers, JWD to Logan, 6 Jan. 1865.
35 C.D. Day to JWD, 16 Jan. 1880.
36 For example, Richard Brown to JWD, 4 Oct. 1869; B.J. Harrington to JWD, 26 Feb. 1871. Joseph Howe to JWD, 23 April 1872.
37 A.R. Selwyn to JWD, 11 Jan. 1870, 29 Oct. 1872.
38 Lyell to JWD, 5 Nov. 1867.
39 For example, JWD to unknown recipient, 8 Dec. 1871.
40 Christopher Dunkin to JWD, 14 Dec. 1869.
41 C.D. Day to JWD, 7 and 15 March 1881.
42 Morris Zaslow, *Reading the Rocks* (Ottawa: Macmillan 1975), 124.

43 Charles MacKineys to JWD, 3 April 1877. Earl of Dufferin to JWD, 20 June 1877; JWD to J.A. Macdonald, Dec. 1879; H. Lyman to JWD, 15 Dec. 1879; JWD to A.R.C. Selwyn, 7 Nov. 1879; Marquis of Lorne to JWD, 19 Feb. 1881; JWD to M.H. Gault, 24 Dec. 1879; JWD to S.L. Tilley, 24 Dec. 1879; T.H.S. White to JWD, 18 Feb. 1881; Thomas Ryan to JWD, 2 March 1881.

44 *Daily Witness*, 28 Feb. 1881; Montreal *Gazette* [?], 28 Feb. 1881; J.F. Whiteaves to JWD, 1 March 1881; Selwyn to JWD, 4 March 1881. For example, JWD to Charles Tupper, 12 Feb. 1881; JWD to T.H.S. White, 17 Feb. 1881.

45 JWD to Lorne, 19 March 1880.

46 JWD to Lyell, 10 Oct. 1868.

47 Dawson, *Fifty Years*, 94. Lyell to Duke of Argyll, 7 March 1868.

48 JWD to Lyell, 18 Jan. 1869.

49 Lyell to Duke of Argyll, 7 March 1868.

50 Lyell to Home, 21 April 1868.

51 Logan to William Chambers, 30 April 1868.

52 D.B. Horn, *A Short History of the University of Edinburgh* (Edinburgh: Edinburgh University Press 1967), 198–9.

53 Logan to JWD, 7 May 1868; EUL: David Masson to William Sharpey, 11 March 1868.

54 Home to JWD, 8 May 1868.

55 J.B. Greenshields to James Alexander, 15 May 1868.

56 JWD to Donald Fraser, 9 May 1868.

57 Fraser to D. Milne Home, 15 May 1868.

58 Lyell to Home, 11 May 1868.

59 Roderick Murchison to JWD, 26 July 1868.

60 The preceding paragraphs draw upon JWD to Chairman of the Board of Curators, 22 May 1868, and "Memorandum with Reference to the Application and Testimonials of Principal Dawson" (MC: Dawson papers).

61 JWD to Anna Dawson, 2 July 1868.

62 "Principalship of the University of Edinburgh," *Scotsman*, 6 July 1868.

63 EUL: James Simpson to Adam Black, 18 June 1868. *The Life of Sir Robert Christison, Bart.*, ed. by his sons, 2 vols. (Edinburgh & London: William Blackwood 1885), vol. 2: 86–8.

64 Peter Bell to JWD, 18, 19 June 1868; Duncan Davidson to JWD, 19 June 1868. Frost, *McGill*, 224.

65 Horn, *Short History*, 200.

66 EUL: David Masson to William Sharpey, 11 March 1868.

67 Christison, *Life*, 92. Horn, *Short History*, 198.

68 EUL: David Masson to William Sharpey, 11 March 1868.

69 Lyell to Duke of Argyll, 7 March 1868; Fraser to Home, 15 May 1868.

70 Lyell to Home, 11 May 1868.

71 Home to Lyell, 3 Aug. 1868.

72 Lyell to JWD, Dec. 1868.
73 JWD to Lyell, 13 Aug. 1868, 18 Jan. 1869.
74 JWD to Duncan Davidson, 28 Sept. 1870.
75 JWD to Lyell, 10 Oct. 1868.
76 See JWD to Lyell, 10 Oct. 1868, where he speaks of public lecturing as providing a pathway to a position in the United States or England.
77 J.J. Bigsby to JWD, 19 Feb. 1872.
78 Bigsby to JWD, 19 Feb. 1872.
79 James McCosh to JWD, 4 Sept. 1872.
80 William Henry Green to JWD, 14 and 23 July 1874.
81 William Adams to JWD, 20 Aug. 1874.
82 E.A. Huntingdon to JWD, 25 Dec. 1880.
83 Jonas Marsh Libbey to JWD, 17 Jan. 1878.
84 Libbey to JWD, 26 Aug. 1879.
85 J.G. Schurman to JWD, 25 July 1881.
86 Libbey to JWD, 16 Dec. 1882.
87 E.J. Craven to JWD, 1 March 1878.
88 McCosh to JWD, 30 Oct. 1876; Craven to JWD, 24 July 1877; George Mercer Dawson to JWD, 23 Aug. 1877.
89 McCosh to JWD, 23 March 1878.
90 W.H. Green to JWD, 27 March 1878.
91 Charles Hodge to JWD, 6 April 1878.
92 A.H. Guyot to JWD, 3 April 1878.
93 McCosh to JWD, 4 April 1878.
94 Peter Redpath to JWD, 5 April 1878.
95 JWD to William Green, 15 April 1878.
96 JWD to Hodge, 15 April 1878.
97 McCosh to JWD, 18 and 25 April, 3 June 1878.
98 T.S. Hunt to JWD, 2 Nov. 1872.

CHAPTER 7

1 Written by another hand on manuscript version of Dawson's autobiography (MC: Dawson papers, box 22, folder 18).
2 Rankine Dawson to G.M. Dawson, 12 Jan. 1900.
3 Cynthia Fish, "Images and Reality of Fatherhood: A Case Study of Montreal's Protestant Middle Class, 1870–1914," Ph.D. dissertation, McGill Dept. of History 1991, abstract and p. 297.
4 MC: Dawson papers, box 64, Anna Dawson, "Notes for a biography of J.W. Dawson," n.d.
5 Anna Dawson, "Notes."
6 MC: George Mercer Dawson papers, fragment entitled "George and his father – Pathetic."
7 For example, JWD to William Dawson, 30 May 1882.

8 MC: Dawson papers, [Clare Harrington?], "Grandmother," 15.
9 Douglas Cole and Bradley Lockner, eds., *The Journals of George M. Dawson: British Columbia,* vol.1: 1875–1878 (Vancouver: University of British Columbia Press 1989), 2–4, 24.
10 G.M. Dawson to JWD, 19 Jan., 15 Feb. 1870.
11 See Cole, *Journals,* 4–7, on George's years at the Royal School of Mines.
12 G.M. Dawson to JWD, 21 Nov. 1869.
13 G.M. Dawson to JWD, 17 Jan. 1872.
14 G.M. Dawson to JWD, 17 Jan., 24 March 1872.
15 G.M. Dawson to JWD, 4 and 26 July 1872.
16 See Cole, *Journals,* 7–8.
17 Cole, *Journals,* 7–8, 25.
18 J.J. Bigsby to JWD, 7 June 1880.
19 See Cole, Journals, esp. 13–18.
20 See Cole, Journals, 18–22.
21 See Bruce Trigger on Dawson's anthropological interests: "Sir John William Dawson: A Faithful Anthropologist," *Anthropologia* 8 (1966): 351–9.
22 JWD to G.M. Dawson, 16 July 1885.
23 G.M. Dawson to JWD, 23 Aug. 1877.
24 G.M. Dawson to JWD, 21 and 24 Oct. 1876. Daniel Wilson to JWD, 15 Dec. 1886. Cole, *Journals,* 25.
25 G.M. Dawson to JWD, 2 Oct. 1878.
26 Anna Dawson to G.M. Dawson, 21 Nov. 1900.
27 MC possesses one folio box (76) of Dawson drawings, including Anna's sketches for *Acadian Geology.* Box 36 (folder 18) holds her drawings for the *Canadian Naturalist.*
28 G.M. Dawson to Anna Dawson, 3 Dec. 1899.
29 Anna Dawson to JWD, 23 April, 2 Nov. 1883.
30 JWD to Anna Dawson, 14 Nov. 1883.
31 B.J. Harrington to JWD, 31 May 1871.
32 Sophie Browne to G.M. Dawson, 8 June 1876. The event would seem to provide excellent evidence for Margaret Westley's arguments about the close connections between all levels of Montreal's "Square Mile" society; see her *Remembrance of Grandeur: the Anglo-Protestant Elite of Montreal, 1900–1950* (Montreal: Libre Expression 1990).
33 It is unclear when the courtship of Anna by Bernard began.
34 B.J. Harrington to JWD, 26 Feb. 1871; 20 June, 13 Dec. 1883.
35 Clare Harrington, "Grandmother," 19, 23.
36 He graduated in arts in 1874 and applied science in 1875.
37 William Dawson to JWD, 18 July 1878; see, for example, William Dawson to JWD, 18, 23, 24 Nov. 1875.
38 William Dawson to JWD, 4 Dec. 1877. Daniel Wilson to JWD, October 1878.

39 William Dawson to JWD, 7 Feb. 1881, 25 March 1878, [1879?].
40 William Dawson to JWD, 3 Sept. 1881.
41 William Dawson to JWD, 26 July, 6 Aug. 1880.
42 Anna Dawson to G.M. Dawson, 18 Feb. 1901.
43 William Dawson to JWD, 7 Feb., 17 Oct. 1881; 8 Feb. 1882.
44 William Dawson to JWD, 6 March 1882.
45 William Dawson to JWD, 14 Feb. 1882, 27 Aug. 1886.
46 William Dawson to JWD, 1 May 1882.
47 Margaret to JWD, 13 April 1886.
48 Cole, *Journals*, 46n.
49 Five lectures on "Is Evolution True?" were published in Sydney by the Christian Workers' Depot.
50 Anna Dawson to B.J. Harrington, May 1886.
51 Sophie Browne to G.M. Dawson, 8 June 1876.
52 Anna Dawson to G.M. Dawson, 21 Nov., 17 Dec. 1900; 25 Feb. 1901.
53 Eva Dawson to G.M. Dawson, 31 Jan. 1901.
54 Rankine Dawson to JWD, 7 July, 10 April 1881; Margaret to JWD, [1881?].
55 Rankine was admitted to the College of Physicians and Surgeons in 1881 (Rankine Dawson to JWD, 22 Oct. 1881). Although Rankine declined to return to "book study" in Montreal, he took a McGill medical degree in 1882 (Margaret to JWD, [1881?]). (According to Winslow-Spragge's "Sketch of My Uncle Rankine, [MC: Dawson papers], this was an honours degree).
56 Rankine Dawson to JWD, 7 and 18 July 1881; 16 Aug. 1883.
57 Rankine Dawson to JWD, 28 Sept. 1881; 11 June, 17 July, 7 Aug. 1882; 10 Sept., 21 Nov. 1885.
58 Cole, *Journals*, 109n.
59 Dawson, Autobiography.
60 J.J. Bigsby, *Thesaurus Devonico-Carboniferous: The Flora and Fauna of the Devonian and Carboniferous Periods. The Genera and Species Arranged in Tabular Form, Showing their Horizons, Recurrences, Localities, and Other Facts. With Large Addenda (from Recent Acquisitions)* (London: John Van Voorst 1878), vii.
61 Dawson, *Some Salient Points in the Science of the Earth* (Montreal: W. Drysdale 1892).
62 J.J. Bigsby, *The Shoe and Canoe or Pictures of Travel in the Canadas Illustrative of their Scenery and Colonial life; with Facts and Opinions on Emigration, State Policy, and Other Points of Public Interest*, 2 vols. (London: Chapman and Hall 1850).
63 Biographical information on Bigsby from DCB and the *Encyclopedia Britannica*.
64 J.J. Bigsby, *Thesaurus Siluricus. The Flora and Fauna of the Silurian Period. With Addenda (from Recent Acquisitions)* (London: John Van Voorst 1868).

65 Bigsby to JWD, 19 Feb. 1872.
66 Bigsby, *Shoe and Canoe*, ix–x.
67 Rankine Dawson, ed., *Fifty Years of Work in Canada, Scientific and Educational* (London and Edinburgh: Ballantyne, Hanson & Co. 1901), 155.
68 Bigsby to JWD, 6 Nov. 1861.
69 I discuss Dawson's palaeobotanical work in chap. 8.
70 Bigsby to JWD, 17 Jan. 1871.
71 Bigsby to JWD, 21 Nov. 1877.
72 Bigsby to JWD, 5 Jan. 1877.
73 Bigsby to JWD, 14 Nov. 1868.
74 Bigsby to JWD, 10 March 1879.
75 For identifications see my "Geological Communication in the Nineteenth Century: The Ellen S. Woodward Autograph Collection at McGill University," *Bulletin of the British Museum (Natural History)*, Historical Series, 10, no. 6 (23 Dec. 1982): 179–226.
76 Bigsby to JWD, 14 Nov. 1868; 23 Sept. 1872; 5 Jan. 1877.
77 For example, Bigsby to JWD, 14 Nov. 1868, 19 Feb. 1872.
78 Bigsby to JWD, 28 June 1870.
79 Bigsby, *Thesaurus Devonico-Carboniferus*, vii.
80 Bigsby, *Thesaurus Siluricus*, vi.
81 Bigsby, *Thesaurus Devonico-Carboniferous*, ix.
82 Bigsby to JWD, 25 April 1871; 5 Jan. 1877, 21 Jan. 1878. In chap. 9, I discuss Dawson's contributions to the periodical.
83 Bigsby to JWD, 5 April 1878.
84 Bigsby to JWD, 5 Dec. 1878.
85 Bigsby to JWD, 11 Nov. 1878.
86 Bigsby to JWD, 16 Oct. 1878.
87 JWD to Lyell, 27 March 1856. Dawson cites Hall's works in his "Fossil Plants from the Devonian Rocks of Canada," *QJGS* 15 (1859): 477–88, and in his "Flora of the Devonian Period in North-Eastern America," *QJGS* 18 (1862): 296–330; see JWD to Lyell, 26 April 1860).
88 See John M. Clarke, *James Hall of Albany: Geologist and Palaeontologist, 1811–1898* (New York: Arno Press 1978 [1st ed. 1923]), 317–23; William E. Eagan, "'I would have sworn my life on your interpretation': James Hall, Sir William Logan, and the 'Quebec Group,'" *Earth Sciences History* 6 (1987): 47–60 (on 51).
89 NYL: JWD to Hall, 10 June 1858.
90 Clarke, *Hall*, 367.
91 For a detailed discussion of this dispute see Eagan, "Quebec Group."
92 Hall to JWD, 24 May 1862.
93 Hall to JWD, 13 Feb. 1862.
94 "James Hall," *DSB*.
95 Hall to JWD, 11 Jan. 1881.

96 These letters are especially valuable because no Wilson correspondence appears to have survived the fire at the University of Toronto.
97 Daniel Wilson to JWD, 20 Dec. 1886.
98 Wilson to JWD, 21 Dec. 1880.
99 See Robert Falconer, "Scottish Influence in the Higher Education of Canada," *Trans. Roy. Soc. Can.* Sect. 2 (1927): 7–20 (on 16). As A.B. McKillop points out in *Matters of Mind: The University in Ontario, 1791–1951* (Toronto: University of Toronto Press 1994), 121, Wilson's brother, George, held the first professorship of technology in the British Empire, at Edinburgh.
100 Wilson to JWD, 19 June 1886.
101 Wilson to JWD, 30 May 1881. Also see McKillop, *Matters of Mind*, 44–5 and 129–32, for Wilson's views on coeducation and politics.
102 Also see McKillop, *Matters of Mind*, for Wilson's commitment to a secular university, 49.
103 Wilson to JWD, 2 Dec. 1880.
104 Wilson to JWD, 17 Jan. 1876.
105 Wilson to JWD, 2 March 1876.
106 Wilson responded to the *Origin of Species* in his presidential address to the Canadian Institute (*Canadian Journal*, Series 2, 5 [1860]: 116–22), but his reaction was openminded and temperate (See Clifford Holland, "First Canadian Critics of Darwin," *Queen's Quarterly*, 88 [1981]: 100–6). Also see McKillop, *Matters of Mind*, 117–19, for Wilson's view of Darwin.

CHAPTER 8

1 Dawson's reference to the rejection of his Bakerian lecture of 1870 by the Royal Society of London's *Phil. Trans.* Much of this chapter appears in a slightly different version in my "'Pearls Before Swine': Sir William Dawson's Bakerian Lecture of 1870," *Notes Rec. R. Soc. Lond.* 45 (1991): 177–91.
2 MC: Dawson papers, box 22, folder 18, Autobiography.
3 Dawson, Autobiography.
4 Dawson, Autobiography.
5 JWD to Lyell, 15 March 1861. Rankine Dawson, ed., *Fifty Years of Work in Canada, Scientific and Educational* (Edinburgh and London: Ballantyne, Hanson & Co. 1901), 137–8.
6 Dawson, Autobiography.
7 JWD to William Williamson, 11 Sept. 1871. Dawson, *Fifty Years*, 140. The Royal Society published only an abstract of Dawson's lecture: "On the Pre-carboniferous Floras of Northeastern America, with special reference to that of the Erian (Devonian) period," *Proceedings of the Royal Society of London* 18 (1869–70): 333–5. (The abstract also appeared in

the *Annals and Magazine of Natural History* 6 [1870]: 103–5). The Geological Survey of Canada subsequently published most of the lecture in two parts (1871 and 1882).

8 With around fifty candidates vying for fifteen available positions, it was rare to be elected within a year, as was Dawson (J.J. Bigsby to JWD, 6 Nov. 1861; Charles Lyell to JWD, 12 June 1861).

9 Lyell to JWD, 12 June 1861. Murchison did recommend Dawson (Murchison to JWD, 10 July 1861). Bigsby later recalled that Dawson's paper had been signed by "six of the greatest geologists in Britain," including Murchison, Leonard Horner, Charles Darwin, and Lyell (Bigsby to JWD, 6 Nov. 1861).

10 As Lyell had warned Dawson (Lyell to JWD, 4 Aug. 1868). Fellows who lived near London and could lobby the administrative committee were most likely to receive these grants, which then totalled only £1000 a year (Roy MacLeod and Peter Collins, eds., *The Parliament of Science: The British Association for the Advancement of Science, 1831–1981* [Northwood, Middlesex: Science Reviews 1981], 103). The Bakerian lectureship had been endowed through a bequest of £100 by Henry Baker in 1774, in order to address a topic in "natural history or experimental philosophy" (Marie Boas Hall, *All Scientists Now: The Royal Society in the Nineteenth Century* [Cambridge: Cambridge University Press 1984], 224 n.13; *The Record of the Royal Society of London*, 1st ed. [London: Harrison & Sons 1897], 124).

11 On the disposition of the royal medals, see Roy MacLeod, "Of Medals and Men: A Reward System in Victorian Science, 1826–1914," *Notes and Records of the Royal Society* 26 (1971): 81–105.

12 Others besides Dawson complained of high fees; Bigsby called the entrance fee "an abominable contrivance to exclude poor and meritorious men" (Bigsby to JWD, 27 April 1861). In 1878, the admission fee of £10 was abolished, and the annual subscription reduced to £3 from £4 (Hall, *All Scientists*, 116–17). Traditionally, the Bakerian lecture paid a fixed sum of £4 to the recipient; this may usefully be compared to the Copley Medal, the highest scientific distinction of the society, which carried a value in gold of about £5 (*Record of the Royal Society* [1897], 124).

13 Since Dawson had opted for the life membership of £60 when he joined the Royal Society (MC: Box 37, memorandum dated 17 July 1862), publication in the *Phil. Trans.* would have reduced the fee to £40, and thus put £20 in his pocket.

14 *The Record of the Royal Society of London*, 3d ed. (London: 1912), 275.

15 William F. Bynum, "Charles Lyell's *Antiquity of Man* and its Critics," *Journal of the History of Biology* 17 (1984): 185.

16 See Bigsby's obituary in the *Proceedings of the Royal Society* 33 (1882): xvi–xvii.

17 William Crawford Williamson, *Reminiscences of a Yorkshire Naturalist*

(London: George Redway 1896), 202–3. This was his "Organisation of the Fossil Plants of the Coal-measures," published in nineteen parts in the *Philosophical Transactions* between 1871 and 1893.
18 Edward Sabine to JWD, 20 April 1870.
19 JWD to Anna Dawson, 26 and 29 April 1870.
20 Sir Joseph Hooker to JWD, 30 April 1870.
21 Logan to JWD, 26 April 1870.
22 Dawson, Autobiography. JWD to Anna Dawson, 26 April 1870.
23 G.M. Dawson to JWD, 28 Oct. 1869.
24 Bigsby to JWD, 12 March 1870.
25 G.M. Dawson to JWD, 8 Dec. 1869; Bigsby to JWD, 12 March 1870; W. Bannister to JWD, 19 May 1870. The Royal Institution lecture, entitled "On the Primitive Vegetation of the Earth" (*Proceedings of the Royal Institution* 6 [1872]: 165–72), was published in modified form in *Nature* 2 (2 June 1870): 85–8 and in the *American Naturalist* 4 (1871): 474–83. The Royal Institution lecture seems to have been arranged by Murchison (G.M. Dawson to JWD, 14 and 28 Oct. 1869).
26 See CUL: J.D. Hooker to C.R. Darwin, 1 Nov. 1862; Darwin to Hooker, 4 and 12 Nov., 24 Dec. 1862. Hooker was referring to his *Outlines of the Distribution of Arctic Plants*, which Dawson reviewed in *CN* 7 (1862): 334–44.
27 Dawson had complained to American botanist Asa Gray that he wished that Hooker's work were "carried on in view of *all the facts*" and "not through the medium of an artificial 'deletion' to support a previous conclusion that species are indefinitely variable." (HUG: JWD to Asa Gray, 19 Sept. 1862.)
28 Hall, *All Scientists*, 115.
29 Bigsby to JWD, 5 Dec. 1878.
30 According to Ruth Barton, this was the period when the X-Club's power began to reach its zenith, due to its domination of council positions ("'An Influential Set of Chaps': The X-Club and Royal Society Politics 1864–85," *British Journal for the History of Science* 23 [1990]: 53–81, esp. 56–9).
31 JWD to Lyell, 12 May 1873.
32 Also see Roy M. MacLeod, "The X-Club: A Social Network of Science in late-Victorian England," *Notes and Records of the Royal Society* 24 (1969): 305–22. In the case of the Geological Society, James Secord showed the considerable power that the Society's president exercised over its publications. See *Controversy in Victorian Geology: the Cambrian-Silurian Dispute* (Princeton, N.J.: Princeton University Press 1986), 231–2.
33 *Proc. Royal Society* 18 (1870): 334.
34 RS: Manuscript of Dawson's Bakerian Lecture, "On the Pre-carboniferous Flora of North-eastern America, and more especially on that of the (Erian) Devonian period," 3, 14. I am indebted to Clelia Pighetti for

loaning me her copy of this manuscript; I have used her pagination in my references.

35 RS: Bakerian Lecture, 8, 98–9, 106.
36 RS: Bakerian Lecture, 9, 100, 105 ff., 108, 119, 106.
37 RS: Bakerian Lecture, 120.
38 "On the Proofs of a Gradual Rising of the Land in Certain Parts of Sweden," *Phil. Trans.* 125 (1835): 1–38.
39 Lyell to JWD, 12 June 1861.
40 Secord, *Controversy,* 232. For other discussions of the avoidance of controversy in early Victorian scientific periodicals see my "From the North to Red Lion Court: The Creation and Early Years of the *Annals of Natural History,*" *Archives of Natural History* 10 (1981): 221–49, and "A Measure of Success: The Publication of Natural History Journals in Early Victorian Britain," *Publishing History* 9 (1981): 21–36. The *American Journal of Science,* in contrast, published only the most controversial section of Dawson's lecture: "On the Bearing of Devonian Botany on Questions as to the Origin and Extinction of Species" 2 (1871): 410–16.
41 RS: Joseph Hooker to George Stokes, 15 May 1870. The sheer length alone of Hooker's report signalled trouble for Dawson; in the Geological Society, for example, most reports were brief (Martin J.S. Rudwick, *The Great Devonian Controversy: The Shaping of Scientific Knowledge among Gentlemanly Specialists* [Chicago: University of Chicago Press 1985], 26).
42 See my *Cathedrals of Science* (Montreal: McGill-Queen's University Press 1988), 33–4, for further remarks on Hooker's criticism of Dawson.
43 P. Martin Duncan (1824–1901), author of *British Fossil Corals,* served as president of the Geological Society in 1877. (Adrian Desmond, *Archetypes and Ancestors: Palaeontology in Victorian London, 1850–1875* [Chicago: University of Chicago Press 1982], 184).
44 JWD to Lyell, 11 June 1860.
45 Desmond, *Archetypes,* 148. See "Joseph Dalton Hooker," DSB 6: 489.
46 RS: "Memo with reference to Principal Dawson's paper on the pre-carboniferous flora," 1 June 1879 (to Mr. White).
47 Walter White to JWD, 8 July 1870.
48 JWD to Lyell, 5 Nov. 1870.
49 William Carruthers to JWD, 4 July 1870.
50 Bigsby to JWD, 23 Sept. 1872.
51 JWD to Lyell, 13 Aug. 1868, 5 Nov. 1870.
52 Henry and Carol Faul, *It Began with a Stone: A History of Geology from the Stone Age to the Age of Plate Tectonics* (New York: John Wiley and Sons 1983), 177.
53 JWD to Lyell, 9 Dec. 1870; JWD to Bigsby, 11 June 1871.
54 Lyell to JWD, 3 Jan. 1871.
55 Bigsby to JWD, 17 Jan. 1871.
56 Bigsby to JWD, 25 April 1871.

57 Bigsby to JWD, 23 Sept. 1872.
58 George Stokes to JWD, 21 March 1871.
59 Rudwick, *Great Devonian*, 26; Secord, *Controversy*, 22–3.
60 RS: JWD to Council, 6 April 1871 (copy in MC).
61 Stokes to JWD, 21 April 1871. The material was also introduced in a course of lectures delivered at the Lowell Institute in Boston in 1887 and eventually published in the "International Scientific Series" as the *Geological History of Plants* (Dawson, *Fifty Years*, 140).
62 JWD to J. Gwyn Jeffreys, 20 Oct. 1871.
63 MC: Dawson papers, box 37.
64 JWD to Jeffreys, 20 Oct. 1871.
65 JWD to Lyell, 27 Oct. 1871.
66 Dawson, Autobiography. JWD to Lyell, April 1871.
67 Fielding Meek to JWD, 16 April 1872. Also see R.W. Ells, "The Progress of Geological Investigation in Nova Scotia," *Proceedings of the Nova Scotia Institute of Science* 10 (1898–1902): 433–46 (on 441). David Penhallow also acclaims Dawson for being the first in North America to study fossil plants by examining the minute anatomy of internal parts ("Canadian Botany from 1800–95," *Trans. Roy. Soc. Can.* ser. 2, 3, Sect. 4 [1897]: 3–56 [on 13]).
68 William C. Darrah, *Principles of Paleobotany* (New York: 1960 [2d ed.]), 55; Faul, *It Began*, 177.
69 T.H. Clark, "Sir John William Dawson (1820–1899) – Paleontologist," *Proceedings of the Geological Association of Canada* 24 (1972): 1–4 (on 2).
70 T.G. Vallance, "The Fuss about Coal: Troubled Relations between Palaeobotany and Geology," *Plants and Man in Australia*, D.J. and S.G.M. Carr, eds. (Sydney: Academic Press 1981), 136–75 (on 139).
71 Desmond, *Archetypes*, 14–15.
72 JWD to J.D. Hooker, 18 Dec. 1878.

CHAPTER 9

1 MC: Dawson papers, box 22, folder 18, Autobiography.
2 Michael Gauvreau describes the evangelical quest to reconcile reason and faith, of which Dawson's works provide an outstanding example; see his *Evangelical Century: College and Creed in English Canada from the Great Revival to the Great Depression* (Montreal and Kingston: McGill-Queen's University Press 1991), 13–14 and elsewhere.
3 EUL: JWD to Lyell, 28 Feb. 1846.
4 Lyell to JWD, 10 August 1846.
5 Katherine M. Lyell, ed., *Life, Letters and Journals of Sir Charles Lyell Bart* (London: John Murray 1881), vol. 2: 181 (Lyell to Leonard Horner, 12 Sept. 1852).
6 JWD to Lyell, Dec. 1847. William Pinnock (1782–1843) published a series

of popular *Catechisms* arranged in the form of questions and answers.
7 Lyell to JWD, 21 April 1856.
8 See Baruch Halpern, *The First Historians: The Hebrew Bible and History* (San Francisco: Harper & Row 1988), 21.
9 In an article entitled "Recent Discussions of the First Chapter of Genesis" (*Expositor* 3d ser., 3 [1886]: 284–301), Dawson wrote that "reasonable men" should attach little importance to the conclusions of the schools of criticism that regard the Pentateuch as of late date, and as made up of several documents (285). One reviewer took Dawson to task for his tone of "superciliousness and contempt" in referring to the German scholars as "bookworms and pedants." (S.R. Driver, "Genesis and Some of its Critics," *Contemporary Review* 55 [1889]: 399–402 [on 402]). Dawson responded that although he had not so impugned the scholars, he did believe that discussion regarding the "antiquity, unity, and genuineness" of Genesis was less interesting than other questions; see "Genesis and Some of Its Critics," *Contemporary Review* 55 (1889): 900–9 (on 901).
10 For example, *Christian World* (Eng.), 23 Nov. 1899.
11 Although H.M. Ami ("Brief Sketch," *Bulletin of the Geological Society of America* 11 [1899]: 550–80) lists an 1857 edition of the book, I have been unable to locate a volume bearing this publication date. Charles F. O'Brien (*Sir William Dawson: A Life in Science and Religion* [Philadelphia: American Philosophical Society 1971]) is inconsistent about the date; Dawson is imprecise in *Fifty Years of Work in Canada, Scientific and Educational* (Rankine Dawson, ed., [London and Edinburgh: Ballantyne, Hanson & Co. 1901]).
12 Dawson Brothers to JWD, 22 March 1875, mentions the publisher's wish to issue a second edition, despite poor sales of the first. The original title proposed by Dawson was *The Mystery of Origins and its Solutions in Revelation and Science* (JWD to Dawson Brothers, 4 Aug. 1876).
13 For example, years later, the secretary of the Royal Society, George Stokes, confused the two works: Stokes to JWD, 21 March 1871.
14 Dawson, *Fifty Years*, 126–7. See DNB entry on John Burnett (1729–84). Burnett had divided his estate between a fund for the poor of Aberdeen, an inoculation endowment, and a trust for the essay (Sumner did win second prize in 1815). The prize was transformed eventually into a lectureship; Stokes delivered the first Burnett lecture in Aberdeen in 1883.
15 JWD to Lyell, Nov. 1859.
16 "James Dwight Dana," DNB vol. 3: 553. YUL: JWD to J.D. Dana, 17 March 1856.
17 Lyell to JWD, 21 April 1856.
18 JWD to Lyell, 11 June 1860.
19 See O'Brien, *Dawson*, 43–58.

20 J.W. Dawson, *Archaia; or, Studies of the Cosmogony and Natural History of the Hebrew Scriptures* (Montreal: B. Dawson & Son; London: Sampson Low, Son & Co. 1860), 41, 48, 128–9.
21 Dawson, *Archaia*, 316–17.
22 Dawson, *Archaia*, 14–15.
23 George L. Parker, *The Beginnings of the Book Trade in Canada* (Toronto: University of Toronto Press 1985), 130, 132. See also his fascinating history of the Dawson publishing firm, (ibid.) 181.
24 Dawson, *Fifty Years*, 128. Most editions bore the imprimatur of Hodder & Stoughton (London) and/or Harper Bros. (New York). An 1882 edition claimed to be the seventh; an 1887 edition, the ninth. The latest I have found was issued in 1903, more than twenty years after the first edition of 1872.
25 Preface to Dawson, *Archaia*.
26 *Bibliotheca Sacra and Biblical Repository* 17 (1860): 443–4.
27 *American Journal of Science* 2d ser., 29 (1860): 146.
28 *Canadian Journal* ser. 2, 5 (1860): 59–62 (on 60).
29 *CN* 4 (1859): 470–93.
30 JWD to Lyell, 11 March 1863.
31 K. Lyell, *Life*, vol. 2: 332: letter from Lyell to JWD of 15 May 1860; JWD to Lyell, 12 Dec. 1866.
32 Henry Acland to JWD, 27 June 1870.
33 JWD to Lyell, 12 May 1873.
34 JWD to Logan, 11 Nov. 1873.
35 Dawson, *The Origin of the World, according to Revelation and Science* (Montreal: Dawson Bros 1877), i. The first edition was published simultaneously in London (Hodder & Stoughton) and New York (Harper Bros).
36 Harper & Bros to JWD, 13 June 1878.
37 JWD to Lyell, 12 May 1873.
38 Dawson, *Origin of the World*, ii–iii.
39 J.J. Bigsby to JWD, 21 Jan., 5 April 1878.
40 E.J. Craven to JWD, 1 March 1878.
41 C. Collingwood to JWD, 4 April 1878.
42 James Smith to JWD, 10 Jan. 1884.
43 J.C. McCulloch to JWD, 7 July 1878.
44 De Sola to JWD, 28 May 1877.
45 O'Brien, *Dawson*, 57.
46 [S.E.D.] [Samuel Edward Dawson], "Nature and the Bible," *CN* ns 8 (1878): 47–54 (on 47, 48).
47 "Nature and the Bible," *CN* ns 8 (1878): 47.
48 J.J. Bigsby to JWD, 25 April 1871; James MacAuley to JWD, 17 Jan. 1871.
49 Henry Anson Buttz to JWD, 22 April 1881.
50 J.R. Pattison to JWD, 22 Dec. 1870.

51 Pattison to JWD, 23 April 1872.
52 J.M. Libbey to JWD, 29 Aug. 1879.
53 A.F. Kemp to JWD, 20 Dec. 1872.
54 James MacAuley to JWD, 17 June 1870; 17 Jan. 1871.
55 JWD to J.R. Pattison, 21 Sept. 1870.
56 MacAuley to JWD, 4 April 1871.
57 MacAuley to JWD, 1 Jan. 1872.
58 Joseph Thomas to JWD, 11 Dec. 1871.
59 NA, v. 363: JWD to John A. Macdonald, 11 Dec. 1879.
60 Hodder & Stoughton to JWD, 6 June 1872; JWD to Hodder & Stoughton, 21 June 1872. Indeed, their view carried the day, as the book was entitled simply *The Story of the Earth and Man*, without even the subtitle proposed by Dawson: "In a series of sketches of the geological periods."
61 JWD to Lyell, 26 April 1860.
62 JWD, "Review of 'Darwin on the Origin of Species by Means of Natural Selection'," *CN* 5 (1860): 100–20 (on 101, 112). Richard Yeo posits that arguments over method always tend to dominate discussion during times of scientific controversy; see "An Idol of the Market-Place: Baconianism in Nineteenth Century Britain," *History of Science* 23 (1985): 251–98.
63 JWD, *Air-Breathers of the Coal Period* (Montreal: Dawson Bros 1873), 77.
64 JWD, "The Present Aspect of Inquiries as to the Introduction of Genera and Species in Geological Time," *The Canadian Monthly* 2 (1872): 154–6 (on 154).
65 [JWD], "Professor Huxley in New York," *International Review* 4 (1877): 34–50 (on 47).
66 JWD, "Review," 116–17; "Present Aspect," 155–6.
67 JWD, "Present Aspect," 154.
68 JWD, "Present Aspect," 113.
69 Peter J. Bowler, *The Eclipse of Darwinism: Anti-Darwinian Evolution Theories in the Decades around 1900* (Baltimore: Johns Hopkins University Press 1983), 44.
70 Bowler, *Eclipse*, 119.
71 MC: Dawson papers, "Recollections of Sir Charles Lyell," 7. [Presidential address to the NHSM, 1875]
72 JWD, "Modern Ideas of Derivation," *CN* ns 4 (1869): 121–38 (on 121–2).
73 JWD to Lyell, 11 June 1860.
74 APS: Darwin to Lyell, 25 June 1860; K. Lyell, *Life*, vol. 2: 333 (Lyell to Darwin, 15 May 1860).
75 Hyatt to JWD, 31 March 1870.
76 Asa Gray, "Sequoia and its History: The Relations of North American to Northeast Asian and to Tertiary Vegetation," Presidential address to the AAAS at Dubuque, Aug. 1872, in his *Darwiniana*, ed. by Hunter Dupree (Cambridge: Belknap Press 1963), 169–94.
77 HUG: JWD to Asa Gray, 8 June 1873.

78 JWD, "Huxley," 36.
79 HUG: JWD to Gray, 8 June 1873.
80 JWD, "Huxley," 36.
81 "The Attitude of Working Naturalists toward Darwinism," reprinted from the *Nation* of 16 Oct. 1873, in Asa Gray, *Darwiniana*, esp. 203, 204.
82 Bigsby to JWD, 5 Jan. 1877.
83 Alexander Winchell to JWD, 6 Sept. 1882; Walcott to JWD, 14 Feb. 1877.
84 Williams to JWD, 6 Jan. 1881.
85 MacAuley to JWD, 4 April, 12 May 1871.
86 As John F. Cornell maintains in "From Creation to Evolution: Sir William Dawson and The Idea of Design in the Nineteenth Century," *Journal of the History of Biology* 16 (1983): 137–70 (139). Cornell's view that the ambiguity of Dawson's later writings on evolution indicates that he could tolerate the theistic evolution of Gray or Mivart is unpersuasive and without foundation in the Dawson correspondence. Just because Dawson gracefully assumed the mantle of elder statesman of science does not mean that he approved of evolutionary doctrines (see 166); his opposition merely became less strident and outspoken.
87 Dawson, *Air-Breathers*, 76. Dawson, "Huxley," 48.
88 Dawson, "Huxley," 48.
89 James C. Livingston, "Darwin, Darwinism, and Theology: Recent Studies," *Religious Studies Review* 8 (1982): 105–16 (on 115).
90 James R. Moore, "Geologists and Interpreters of Genesis in the Nineteenth Century," in David Lindberg and Ronald Numbers, eds., *God and Nature: Historical Essays on the Encounter between Christianity and Science* (Berkeley: University of California Press 1986), 322–50 (on 340–1).
91 *New York Commercial Advertiser*, 20 Nov. 1899.
92 *The Outlook* (London), 25 Nov. 1899.
93 Margaret to JWD, 1886.

CHAPTER 10

1 Charles F. O'Brien, *Sir William Dawson: A Life in Science and Religion* (Philadelphia: American Philosophical Society 1971), 115.
2 JWD to Lyell, Aug. 1864.
3 O'Brien, *Dawson*, 182.
4 Rankine Dawson, ed., *Fifty Years of Work in Canada, Scientific and Educational* (London and Edinburgh: Ballantyne, Hanson & Co. 1901), 144.
5 JWD to Lyell, 5 Jan. 1866.
6 MC: Dawson papers, box 22, folder 18, Autobiography, and *Fifty Years*, 136.
7 J.D. Dana to JWD, 24 June 1872.
8 O'Brien, *Dawson*, 37, 90.
9 As Richard Yeo suggests, methodological criticisms function as "argumentative devices" in scientific controversies. ("An Idol of the Market-

Place: Baconianism in Nineteenth Century Britain," *History of Science* 23 [1985]: 251–98.)
10 Dawson, "On Mr. Carter's Objections to *Eozoön*," *Annals of Natural History* ser. 4, 17 (1876): 118–19.
11 JWD to Lyell, 20 Oct. 1865.
12 Dawson to Asa Gray, 19 Sept. 1862
13 Dawson, Autobiography.
14 For an extended discussion of "the glacier question" and Dawson's role therein, see O'Brien, *Dawson*, chap. 7. George P. Merrill also devotes a chapter to glacial theory in his *First One Hundred Years of American Geology* (New York and London: Hafner Publishing Co. 1964).
15 O'Brien, *Dawson*, 152.
16 O'Brien, *Dawson*, 160.
17 O'Brien, *Dawson*, 153–4.
18 *CN* ns 1 (1864): 221ff.; JWD to Lyell, 20 Oct. 1865.
19 JWD to Lyell, 28 April 1864.
20 O'Brien, *Dawson*, 160–3.
21 See extended discussion in O'Brien, Dawson, 164–9.
22 See Sheets-Pyenson, *Index to the Scientific Correspondence of John William Dawson* (BSHS Monograph 7) (Stanford in the Vale: British Society for the History of Science 1992) for relevant letters.
23 S.A. Miller to JWD, 12 April 1878.
24 See O'Brien, *Dawson*, 170–5; J.W. Spencer to JWD, 25 May 1887.
25 Merrill, *American Geology*, discusses the connection between the Noachian deluge idea and glacial theory; esp. see 624–5. Martin Rudwick suggests that William Buckland's identification of "the geological 'diluvial' episode with the biblical 'Flood'" was not typical ("The Glacial Theory," *History of Science* 8 [1969]: 136–57 [on 138–9]). For a witty and engaging account of the glacier controversy see A. Hallam, *Great Geological Controversies* (Oxford: Oxford University Press 1989 [2d ed.]), 87–104.
26 Lorne to JWD, 26 Aug. 1880.
27 Prestwich to JWD, 29 July 1892.
28 Dawson, *The Canadian Ice Age: Being Notes on the Pleistocene Geology of Canada, with Especial Reference to the Life of the Period and its Climatal Conditions* (Montreal: William V. Dawson 1893), 289. JWD, "Presidential Address to the NHSM," *CRS* 3 [1889]: 436–40 [on 438]).
29 BMS: JWD to Scudder, 14 July 1881.
30 *CRS* 3: 438.
31 JWD to [Lewis], 24 Aug. 1881; William E. Eagan, "The Multiple Glaciation Debate: The Canadian Perspective, 1880–1900," *Earth Sciences History* 5 (1986): 144–51.
32 Dawson, *Ice Age*, vi.
33 Dawson, *Ice Age*, 287.
34 Dawson, *Ice Age*, 287–9.

35 In one of his addresses to the NHSM, Dawson referred to *Eozoön* as being the brightest jewel in the crown of the Geological Survey of Canada; see Charles F. O'Brien, "*Eozoön Canadense* 'The Dawn Animal of Canada'," *Isis* 61 (1970): 206–25 (on 208), and Merrill, *American Geology*, 568.
36 For a detailed discussion of the controversy and Dawson's role therein, see O'Brien, "*Eozoön*," esp. 207–8. Also see Merrill, American Geology, chap. 10.
37 T.H. Clark, "Informal interview with P.R. Eakins and students, 1971." (Audiotape supplied courtesy of Ingrid Birker, curator of Paleontology, Peter Redpath Museum)
38 O'Brien, *Dawson*, 208–9.
39 JWD to Lyell, 5 Jan., 29 March 1866.
40 O'Brien, *Dawson*, 209.
41 Lyell to JWD, 11 July 1867, 18 Feb. 1868.
42 JWD to Lyell, 5 Jan. 1866.
43 See O'Brien, "*Eozoön*," 210 ff., for a detailed discussion.
44 JWD to Lyell, 20 Jan. 1871.
45 K.H.G. Credner to JWD, 29 Nov. 1868.
46 John Plant to JWD, 1 March 1871. Individuals seeking *Eozoön* specimens account for hundreds of letters in the Dawson correspondence at MC.
47 R. Damon to JWD, 9 April 1879.
48 T.S. Hunt to JWD, 19 June 1876. Edward [?] Gaupen to JWD, 25 Sept. 1880.
49 Samuel Haughton to JWD, 9 Feb. 1871.
50 H.B. Brady, 12 Feb. 1871.
51 See O'Brien, "*Eozoön*," for details, 208–16.
52 Presumably, this refers to the London optical instrument firm Richard and Joseph Beck, maker of light microscopes.
53 Thomas Rupert Jones (1819–1911), professor of geology at the Royal Military College and at the Staff College, Sandhurst.
54 (1832–1913), professor of comparative anatomy, University of Prague, and director of the Natural History Department, Zoological Museum of Bohemia.
55 Sir Andrew Crombie Ramsay (1814–91), director general of the Geological Survey of Great Britain.
56 A newspaper clipping of this poem is filed in box 35 of the Dawson papers (MC); no attribution or identification is supplied.
57 MC: Logan papers, JWD to Logan, 18 April 1874. *The Dawn of Life; being the History of the Oldest Known Fossil Remains, and their relations to geological time and to the development of the animal kingdom* (London: Hodder & Stoughton 1875).
58 Dawson, *Dawn*, Preface and "introductory," vii, 2, 4.
59 JWD to Logan, 18 July 1874. Harper & Co. in New York also deduced that this would be the case; see them to JWD, 30 Nov. 1875.
60 Dawson, *Dawn*, 4–5.

61 Hodder & Stoughton to JWD, 20 Jan. 1876.
62 Hodder & Stoughton to JWD, 18 April 1878.
63 O'Brien, "*Eozoön*," 217.
64 W.H. Brock and A.J. Meadows, *The Lamp of Learning: Taylor & Francis and the Development of Science Publishing* (London: Taylor & Francis 1984), 114.
65 O'Brien, "*Eozoön*," 215–16. JWD to A.C. Ramsay, 5 March 1875.
66 JWD to Ramsay, 5 March 1875.
67 O'Brien, "*Eozoön*," 217.
68 JWD to W.B. Carpenter, 24 March 1876.
69 The first was not really a "review," but a comment by H.J. Carter: "Relation of the Canal-System to the Tubulation in the Foraminifera, with reference to Dr. Dawson's 'Dawn of Life'," *Annals of Natural History* ser. 4, 16 (1875): 420–4. The "second" review – the actual book review – was written by Rowney and King and appeared in 17 (1876): 360–77.
70 *Annals of Natural History* ser 4, 17 (1876): 371.
71 JWD to Dallas, 27 May 1876.
72 JWD to Dallas, 1 May 1876.
73 J.D. Dana to JWD, 19 April 1876.
74 W.B. Carpenter to JWD, 16 May 1876.
75 Carpenter to JWD, 16 May 1876.
76 Carpenter to JWD, 16 May 1876.
77 As H.B. Brady suggests in his letter to Dawson of 5 Sept. 1879.
78 William B. Dawson to JWD, 5 Nov. 1877.
79 G.M. Dawson to JWD, 30 Aug. 1877.
80 JWD to W.B. Carpenter, 7 Dec. 1877.
81 R. Rathbun to JWD, 27 Feb. 1879.
82 W.O. Crosby to JWD, 7 Oct. 1878.
83 O'Brien, "*Eozoön*," 219.
84 O'Brien, "*Eozoön*," 220.
85 O'Brien, "*Eozoön*," 220–1.
86 J.D. Dana to JWD, 23 Jan. 1880.
87 P.H. Carpenter to JWD, 23 Nov. 1885, 3 Jan. 1886; JWD to Carpenter, 18 Jan. 1886.
88 O'Brien, "*Eozoön*," 221.
89 Dawson, "Review of the Evidence for the Animal Nature of *Eozoön* Canadense," CRS 6 (1895): 470 (reprinted from *Geological Magazine*, Decade IV, 2 [Oct-Dec. 1895]).
90 O'Brien, "*Eozoön*," 221–3.

CHAPTER 11

1 See, for example, his obituary in *Engineering and Mining Journal* (NY), 25 Nov. 1899.

2 Rankine Dawson, ed., *Fifty Years of Work in Canada, Scientific and Educational* (Edinburgh and London: Ballantyne, Hanson & Co. 1901), 110.
3 JWD to Lyell, Aug. 1864.
4 HUM: JWD to H.A. Morrell, 10 Oct. 1870.
5 See William A.S. Sarjeant and David J. Mossman, "Vertebrate Footprints from the Carboniferous Sediments of Nova Scotia: A Historical Review and Description of Newly Discovered Forms," *Palaeogeography, Palaeoclimatology, Palaeoecology* 23 (1978): 279–306 (on 280–3). These two contributions were "On the Lower Carboniferous Rocks, or Gypsiferous Formation of Nova Scotia" (*Proc. Geol. Soc. Lond.* [1843]: 272–81) and "On the Newer Coal Formation of the Eastern Part of Nova Scotia" (*Proc. Geol. Soc. Lond.* [1845]: 504–12).
6 Richard Owen, "Description of Specimens of Fossil Reptilia Discovered in the Coal-Measures of the South Joggins, Nova Scotia, by Dr J.W. Dawson, F.G.S.," *QJGS* 18 (1862): 238–44.
7 ICS: Lyell to T.H. Huxley, 13 Feb. 1863.
8 NHM: JWD to Richard Owen, 24 Oct., 8 Dec. 1862; 30 Jan. 1863.
9 See Dawson's response: "Notice of a New Species of Dendrerpeton, and of the Dermal Coverings of Certain Carboniferous Reptiles, *QJGS* 19 (1863): 469–73.
10 ICS: Lyell to Huxley, 13 Feb. 1863.
11 ICS: Lyell to Huxley, 28 Nov. 1863.
12 ICS: Copies of letters, Huxley to Lyell, n.d. [1866?].
13 *CN* 8 (1873): 1–12, 81–92, 159–60, 161–75, 268–95.
14 JWD to Lyell, 26 April 1860.
15 "On the Results of Recent Explorations of Erect Trees containing Animal Remains in the Coal-formation of Nova Scotia," *Phil. Trans.* 173, pt. 2 (1882): 621–59.
16 Sargeant, "Footprints": 282, 284–88. This was his "Synopsis of the Air-breathing Animals of the Palaeozoic in Canada, up to 1894," *Trans. R. Soc. Can.* 12, sect. IV (1895): 71–88.
17 NYL: JWD to James Hall, 10 June 1858.
18 Dawson, *Supplementary Chapter to Acadian Geology* (Edinburgh: Oliver & Boyd 1860), 1.
19 *CN* 5 (1860): 450–5.
20 And simultaneously by Oliver & Boyd in Edinburgh; A. & W. Mackinlay in Halifax, and Dawson Bros. in Montreal.
21 Dawson, *Acadian Geology. The Geological Structure, Organic Remains, and Mineral Resources of Nova Scotia, New Brunswick, and Prince Edward Island* (London: Macmillan; Edinburgh: Oliver & Boyd; Montreal: Dawson Brothers 1878 [3d ed.]), 7.
22 MC: Dawson papers, box 27, folder 39, prospectus entitled "Second Edition of *Acadian Geology*," Sept. 1866.

23 Dawson, *Acadian Geology. The Geological Structure, Organic Remains, and Mineral Resources of Nova Scotia, New Brunswick, and Prince Edward Island* (London: Macmillan; Edinburgh: Oliver & Boyd; Halifax: A. & W. Mackinlay; Montreal: Dawson Brothers 1868) [2d ed.], viii, 421.
24 JWD to Lyell, 12 Dec. 1866. Also see printed prospectus for the second edition, with notations by Dawson (MC: Dawson papers, box 27, folder 39).
25 Dawson, *Acadian Geology*, 2d ed., "preface," vi.
26 *Acadian Geology*, 2d ed., 4.
27 *Acadian Geology*, 2d ed., 66.
28 *Acadian Geology*, 2d ed., 202 ff., 225 ff.
29 *Acadian Geology*, 2d ed., 396, 412.
30 Dawson, *Acadian Geology*, 2d ed., 497. On these controversies see Martin J.S. Rudwick, *The Great Devonian Controversy* (Chicago: University of Chicago Press 1985) and James A. Secord, *Controversy in Victorian Geology* (Princeton: Princeton University Press 1986).
31 Dawson, *Acadian Geology*, 3d ed., 78.
32 Dawson, *Acadian Geology*, 2d ed., 523.
33 Dawson, *Acadian Geology*, 2d ed., 531–2.
34 Dawson, *Acadian Geology*, 2d ed., 557, intro to chap. 23.
35 Dawson, *Acadian Geology*, 2d ed., 638.
36 Dawson, *Acadian Geology*, 2d ed., 658–9.
37 Dawson, *Acadian Geology*, 2d ed., 664–8.
38 Dawson, *Acadian Geology*, 2d ed., 668–71.
39 Dawson, *Acadian Geology*, 2d ed., 669–70.
40 Lyell to JWD, 5 Nov. 1867.
41 Katherine M. Lyell, ed., *Life, Letters and Journals of Sir Charles Lyell Bart* (London: John Murray 1881), vol. 2: 428.
42 Lyell to JWD, 4 Aug. 1868.
43 Lyell to JWD, 1 July 1868.
44 Lyell to JWD, 1 July 1868.
45 For example, see JWD to John Phillips, 13 April 1868.
46 MC: Dawson papers, box 37 (Scrapbook), printed circular.
47 Bourinot to JWD, 11 Sept. 1869; JWD to Drummond Campbell, 28 Feb. 1870.
48 L.W. Bailey to JWD, 23 April 1867.
49 MC: Dawson papers, box 37, printed circular.
50 "Sketch of Principal Dawson," reprinted from the *Popular Science Monthly* (1875): 1–4 (on 4).
51 *CN* ns 8 (1878): 472–5; *American Journal of Science* ser. 3, 15 1878: 478–80. Both of these notices seem to have been prepared by Dawson.
52 Dawson, *Acadian Geology*, 3d ed., 7–8.
53 Dawson, *Acadian Geology*, 3d ed., 8.
54 George P. Merrill discusses the third edition of *Acadian Geology* on pp.

489–91 of *First One Hundred Years of American Geology* (New York and London: Hafner Publishing 1964).
55 G.F. Matthew to JWD, 8 April 1878.
56 Oliver & Boyd to JWD, 24 Feb. 1879; Alexander Macmillan to JWD, 16 April 1879.
57 Macmillan to JWD, 20 Nov. 1879; Oliver & Boyd to JWD, 24 May 1884.
58 JWD to Oliver & Boyd, 30 July 1880. This was his *Lecture Notes on Geology and Outline of the Geology of Canada for the Use of Students* (Montreal: Dawson Bros. 1880).
59 Dawson, *The Geology of Nova Scotia, New Brunswick, and Prince Edward Island, or Acadian Geology* (London: Macmillan & Co. 1891 [4th ed.]), 3.
60 Dawson, *Geology of Nova Scotia*, 5.
61 Dawson, *Geology of Nova Scotia*, 6, 8.
62 Dawson, *Geology of Nova Scotia*, 37.
63 See Merrill, *American Geology*, 359.
64 Dawson, *Geology of Nova Scotia*, 13–14.
65 Dawson, *Fifty Years*, 142.
66 R. Brown to JWD, 18 May 1869.
67 James Drummond to JWD, 4 May 1869.
68 *Pictou Advocate*, 11 Oct. 1866.
69 James M. Cameron, *The Pictonian Colliers* (Halifax: Nova Scotia Museum 1974), 33.
70 John Rutherford, *The Coal-Fields of Nova Scotia* (Newcastle-upon-Tyne: A. Reid 1871), 4.
71 Although according to Richard Brown, Cape Breton bituminous coal had already been losing out to Pennsylvania anthracite in the United States (*Coal Fields*, 94 ff.).
72 Brown to JWD, 2 April 1869.
73 Henry Budden to JWD, 21 Sept. 1871.
74 Drummond Campbell to JWD, 22 Feb. 1870; JWD to Campbell, 28 Feb. 1870.
75 G.H. Dobson to JWD, 23 Feb. 1877; JWD to Dobson, 27 Feb. 1877.
76 See Richard Brown, *Coal Fields*; for example, 155–6.
77 R. Haliburton to JWD, 1 July 1872.
78 E.A. Prentice to JWD, 2 June 1870, 6 June 1872.
79 JWD to Prentice, 1 July 1872; Prentice to JWD, 6 July 1872.
80 "Memo on Iron Ores, Pictou" dated August 1872, Box 26 (MC: Dawson papers).
81 Prentice to JWD, 4 July 1872.
82 E. Primrose to JWD, 24 July 1872.
83 G.M. Dawson to JWD, 4, 26 July 1872.
84 G.M. Dawson to JWD, 12 Sept. 1872.
85 JWD to G.M. Dawson, 5 Oct. 1872.

86 Thomas Dickson to JWD, 14 Oct. 1872; JWD to Alfred Maddick & Co., 7 March 1873; JWD to Josiah Deacon, 30 Oct. 1872.
87 Howard Primrose to JWD, 12 Dec. 1872.
88 Howard Primrose to JWD, 5 Dec. 1872.
89 JWD to G.M. Dawson, 5 Oct. 1872.
90 "Nova Scotia Iron Manufacture," *Gazette*, 7 Jan. 1873.
91 Howard Primrose to JWD, 25 March 1873.
92 Howard Primrose to JWD, 2 June 1876.
93 JWD to Peter Redpath, June 1876; Redpath to JWD, 5 June 1875.
94 Anna Grant to JWD, 13 March 1879.
95 W.J. Ross to JWD, 19 March 1879.
96 According to Richard Brown (*Coal Fields*, 110), gold mining never yielded a profit in Nova Scotia.
97 Joseph Howe to JWD, 30 July 1861.
98 *Free Press* [Winnipeg], 20 Nov. 1899.

CHAPTER 12

1 *The Spectator*, 18 May 1878. For a good discussion of the poem's allusions see Edward Andrew Collard, *All Our Yesterdays* (Montreal: *The Gazette* 1988), 110–11.
2 The archival records of the Natural History Society of Montreal appear in a number of volumes housed in the Rare Book Collection of the Blacker-Wood Library at McGill. These include manuscript minute books and records of council proceedings, as well as a printed series of annual reports. Published proceedings of the society also appear in the *Canadian Naturalist*, and its successor the *Canadian Record of Science*. According to the minutes, Dawson was elected to the society on 31 Dec. 1855, just several months after his arrival in Montreal.
3 For a superb analysis of the functions and meanings of amateur natural history societies in Canada, among them the NHSM, see Carl Berger, *Science, God, and Nature in Victorian Canada* (Toronto: University of Toronto Press 1983), chap. 1. Also see Suzanne Zeller, *Inventing Canada: Early Victorian Science and the Idea of a Transcontinental Nation* (Toronto: University of Toronto Press 1987), 31ff, for a history of the NHSM.
4 Berger, *Science*, 8.
5 MBW: NHSM Minutes, 26 May 1856.
6 MBW: NHSM Annual Report for 1860, 32.
7 MBW: NHSM minutes, 24 Nov. 1856, 382–3.
8 MBW: NHSM minutes, 24 Nov. 1856, 381. Stanley Brice Frost, "Science Education in the Nineteenth Century: The Natural History Society of Montreal, 1827–1925," *McGill Journal of Education* 17 (1982): 31–43 (on 40–1). Another brief discussion of the NHSM appears in Peter J. Bowler,

"The Early Development of Scientific Societies in Canada," in A. Oleson and S.C. Brown, eds., *The Pursuit of Knowledge in the Early American Republic* (Baltimore: Johns Hopkins University Press 1976), 328–9 (on 326–9). *CN* 3 (1858): 399.
9 Dawson, "Things to be observed in Canada, and especially in Montreal and its vicinity," *CN* 3 (1858): 1–12.
10 Asa Gray to JWD, 5 July 1856.
11 *CRS* 4 (1891): 70.
12 MBW: NHSM, Minutes of the annual general meeting of 1858, vol. 1.
13 MBW: NHSM, Proceedings of 1873, 15.
14 *CRS* 3 (1889): 439–40.
15 *CN* ns 9 (1881): 374.
16 MBW: NHSM, Minute book, printed report for 1857, 401–2. Minutes of 25 May and 29 June 1857.
17 "Eleventh Meeting of the American Association for the Advancement of Science," *CN* 2 (1857): 241–99 (on 242).
18 Sally Gregory Kohlstedt, *The Formation of the American Scientific Community: The American Association for the Advancement of Science, 1848–60* (Urbana: University of Illinois Press 1976), 201, table 7 (although Kohlstedt cautions that her figures are only approximate, and that attendance at these two meetings was artificially inflated by the local populace who joined the AAAS for these years only, 195).
19 See Kohlstedt, *Formation*, and *CN* ns 9 (1881): 374.
20 "Further Gleanings from the Meeting of the American Association in Montreal," *CN* 2 (1857): 356.
21 *CN* 2 (1857): 298–9; 3: 228–9.
22 Mary Lesley Ames, ed., *Life and Letters of Peter and Susan Lesley* (New York and London: Knickerbocker Press, 1909 2 vols.), vol. 1: 350 [Peter Lesley to his wife, 13 Aug. 1857].
23 *CN* 3 (1858): 228–9.
24 E. Billings to JWD, October 1856.
25 For a short history of the changes in title, periodicity, and editors of the *Canadian Naturalist*, see its successor: *Canadian Record of Science* 1 (1885): i–ii.
26 MBW: NHSM, Annual Report for 1859, 9. *CN* 5 (1860): 233.
27 *CN* ns 4 (1869): 212.
28 *CN* ns 5 (1870): 215–16.
29 Dawson Brothers to JWD, 24 Feb. 1872.
30 B.J. Harrington to JWD, 13 Dec. 1883.
31 *CN* 5 (1860): 237.
32 H.A. Ward to JWD, 25 April 1869.
33 Victor Lyon to JWD, 28 Jan. 1879.
34 *CN* ns 5 (1870): 208.

35 *CN* 5 (1860): 232.
36 *CRS* 3 (1889): 434, 437.
37 *CRS* 5 (1893): 196.
38 *CN* 8 (1863): 217.
39 *CN* ns 6 (1872): 8.
40 *CN* ns 3 (1866): 132.
41 JWD to Samuel Scudder, 6 Oct. 1869.
42 G.M. Dawson to JWD, 6 Nov. 1870.
43 H.A. Nicholson to JWD, 17 Feb. 1872.
44 *CN* ns 4 (1869): Dawson, "Modern Ideas of Derivation," 121–38 (on 121).
45 For example, *CN* ns 1 (1864): 220–8; ns 7 (1875): 281.
46 James Hall to JWD, 20 Jan. 1859. SIA: JWD to Spencer Baird, 24 Oct. 1859; JWD to H.A.J. Verreau, 17 Feb. 1871.
47 JWD to Robert Bell, 20 Feb. 1863.
48 Gerard E. Hart to JWD, 4 June 1877; J.B. Edwards to JWD, 23 Jan. 1878.
49 *CN* 3 (1858): 230, 399.
50 *CN* 5 (1860): 230.
51 *CN* 7 (1862): 229–30.
52 *CN* ns 2 (1865): 76.
53 *CN* ns 1 (1864): 57–8, 64.
54 *CN* ns 2 (1865): 305.
55 *CN* ns 3 (1868): 395.
56 *CN* ns 4 (1869): 66.
57 *CN* ns 5 (1870): 80–1.
58 *Star,* 12 Jan. 1887.
59 *Gazette,* 17 Jan., 21 Jan. 1887; clippings in Minutes of the Conversazione Committee, 18, 25 (MBW).
60 *CN* ns 3 (1868): 396.
61 *CRS* 3 (1889): 180–1.
62 *CRS* 3 (1889): 432.
63 *CRS* 3 (1889): 442–3.
64 *CN* ns 2 (1865): 303.
65 *CN* ns 1 (1864): 52.
66 *CN* 7 (1862): 232–3; *CN* ns 1 (1864): 51.
67 *CN* ns 3 (1868): 396, 398.
68 *CN* ns 4 (1869): 211; *CN* ns 5 (1870): 80.
69 MBW: NHSM, Annual reports for 1859 (10), 1871 (10); Minutes, vol. 2 [for 1889?].
70 *CN* ns 3 (1868): 392.
71 MBW: NHSM, Proceedings of 1871, 10.
72 *CN* ns 4 (1869): 211.
73 *CN* ns 8 (1878): 179; ns 9 (1881): 381.
74 *CN* ns 7 (1875): 108.

75 Selwyn told Dawson not to even "entertain" the idea of electing him president; others besides the two of them should serve (A.R.C. Selwyn to JWD, 27 April 1871).
76 *CN* ns 8 (1878): 449–50.
77 *CN* ns 9 (1881): 384.
78 J.F. Whiteaves to JWD, 2 Sept. 1876.
79 J.F. Whiteaves to JWD, 17 June 1870; JWD to Christopher Dunkin, 3 May 1871.
80 *CN* ns 7 (1875): 293; *CN* ns 9 (1881): 181; J.G. Robertson to JWD, 21 May 1885; T.S. Hunt to JWD, 26 March 1885; W.W. Lynch to JWD, 8 June 1885; 26 April 1886.
81 See Berger, *Science,* and Zeller, *Inventing,* on the relationship between the NHSM and the Geological Survey of Canada. JWD to James Ferrier, 17 March 1877.
82 JWD to Ferrier, 17 March 1877; JWD to Thomas Ryan, Feb. 1881; Ryan to JWD, 2 March 1881; A.R.C. Selwyn to JWD, 4 March 1881.
83 *CN* ns 9 (1881): 475.
84 *CN* ns 10 (1883): 107.
85 *CN* ns 8 (1878): 450.
86 *CN* ns 10 (1883): 241.
87 Vittorio M.G. de Vecchi, "The Dawning of a National Scientific Community in Canada, 1878–1896," *Scientia Canadensis* 26 (1984): 32–58 (on 42).
88 JWD to unknown correspondent, 1880.
89 *CN* ns 9 (1881): 440.
90 *CN* ns 10 (1883): 110.
91 *CN* ns 10 (1883): 242–3.
92 *CRS* 5 (1893): 198.
93 *CRS* 3 (1889): 445–6; *CN* ns 10 (1883): 109–10.
94 *CN* ns 10 (1883): 245; *CRS* 3 (1889): 441.
95 *CRS* 6 (1895): 103–4.
96 *CRS* 8 (1899): 247.

CHAPTER 13

1 Slightly different versions of portions of this chapter were published earlier in "Better than a Travelling Circus: Museums and Meetings in Montreal during the Early 1880s," *Transactions of the Royal Society of Canada,* Ser. 4, 20 (1982): 599–618.
2 F.W. Putnam to JWD, 8 April 1880.
3 JWD to Putnam, 10 June 1880.
4 Putnam to JWD, 20 Aug. 1881.
5 HUA: JWD to Putnam, 22 Dec. 1881.
6 *CN* ns 10 (1883): 60.

7 Peter Redpath to JWD, 10 March 1882.
8 F.W. Hicks to JWD, 2 Aug. 1882; A.H. Winchell to JWD, 27 Jan. 1883.
9 Daniel Wilson to JWD, 20 April 1882.
10 M.M. Dawson to JWD, July 1882.
11 M.M. Dawson to JWD, 1882.
12 Alexander Johnson to JWD, 25 July 1882.
13 Rankine Dawson to JWD, 7 Aug. 1882.
14 Rankine Dawson to JWD, 27 Aug. 1882.
15 B.J. Harrington to JWD, 16 Aug. 1882.
16 HUM: JWD to Alexander Agassiz, 10 July 1882.
17 Putnam to JWD, 10 Aug. 1882.
18 M.M. Dawson to JWD, 1882; HUG: JWD to Asa Gray, 15 Aug. 1882.
19 MC: Dawson papers, box 37, *Witness*, 25 Aug. 1882; press clipping of 25 Aug. 1882. JWD to W. Douw Lighthall, 28 Aug. 1882.
20 Peter Redpath to JWD, 10 March, 25 April 1882.
21 J.S. Copes to JWD, 3 July 1882; Benjamin Silliman to JWD, 12 Aug. 1882.
22 Leo Lesquereux to JWD, 25 Oct. 1882.
23 Gray to JWD, 7 June 1882; Putnam to JWD, 7 June 1882.
24 *Proceedings of the American Association for the Advancement of Science, 31st meeting, Montreal* (Salem: 1883), 623; MC: Dawson papers, box 37, press clippings of 24 Aug. 1882.
25 A.E. Foote to JWD, 24 July 1882; Henry Ward to JWD, 6 July 1882; HUM: JWD to Alexander Agassiz, 18 July 1882.
26 *AAAS Proc.*, 635; "The American Association for the Advancement of Science," *CN* ns 10 (1883): 378–84 (on 384).
27 J.F. Torrance to JWD, 23 Jan. 1882.
28 J.W. Judd to JWD, 19 May 1881; T.R. Jones to JWD, 12 May 1881; H. Hicks to JWD, 25 July 1882; G.M. Dawson to JWD, 20 June 1882; "Science in America," *Leisure Hour* (1882): 512; *AAAS Proc.*, 635.
29 MBW has two boxes and two scrapbooks that deal with the 1884 British Association meeting. These materials were originally collected by the Montreal Natural History Society. See "Circular," 8 Nov. 1882.
30 John Rae to JWD, 12 April 1883.
31 "The Proposed Meeting of the British Association for the Advancement of Science in Canada," *Science*, 4 May 1883.
32 *CRS* 7 (1897): 397.
33 O.J.R. Howarth, *The British Association for the Advancement of Science: A Retrospect, 1831–1921* (London: British Association 1922), 120.
34 Lorne to JWD, 27 Feb. 1883.
35 Samuel Scudder to JWD, 7 April 1883; "The Proposed Meeting of the British Association for the Advancement of Science in Canada," *Science* 1 (May 1883): 351–2. As early as 1881, the AAAS sent a delegation to York to propose an International Scientific Association, which was rejected as

chimerical by the British body ("The British Association," *Gazette*, 27 Oct. 1882). Apropos the growing sentiment for an international meeting, see also AAAS *Proc.*, 623–4.

36 Joseph Colmer to JWD, 6 Dec. 1883.

37 MBW: British Association materials, "The British Association," *Gazette*, 27 Oct. 1882; "The British Association," *Montreal Herald*, 23 Dec. 1882; "The British Association," *Gazette*, 11 Oct. 1883.

38 Daniel Wilson to JWD, 3 Sept. 1883; Anna Dawson to JWD, Nov. 1883; see also account in the *Gazette*, 11 Oct. 1883; T.S. Hunt to JWD, 3 Dec. 1883; B.J. Harrington to JWD, 13 Dec. 1883.

39 Peter Redpath to JWD, 5 Feb. 1884; "The British Association," *Herald*, 3 Dec. 1883.

40 "The British Association," *Herald*, 11 Jan. 1884; see also BAAS, Montreal Meeting, 1884, *Canadian Economics, being papers prepared for reading before the economical section* (Montreal: Dawson Brothers 1885), "Preliminary arrangements and meeting in Canada," xxi–xxxi.

Robert Moat replaced F.W. Thomas as chairman of the finance committee; R.R. Grindley, in turn, replaced Moat when he left for England in July 1884. George Drummond, who succeeded Hugh McLennan as head of the hospitality committee, resigned around the same time; he was replaced by Thomas Workman. Secretaries also changed: F.S. Lyman replaced J.S. McLennan on the reception committee; Alexander Robertson succeeded McDonnell on "meeting places."

41 MC: Dawson papers, box 35, "The British Association in Canada," read by Henry Lefroy to the Royal Colonial Institute, 13 Jan. 1885, 3.

42 JWD to G.M. Dawson, 21 July 1884.

43 Joseph Colmer to Thomas White, 2 Oct. 1883 (MBW: large scrapbook, 9; circular dated 1 March 1884 from the Citizens' Executive Committee).

44 "The British Association and the French Population," *Montreal Daily Star*, 28 May 1884; "The British Association," *Herald*, [late August] 1884 (MBW: Large scrapbook, 83).

45 *The British Association's Visit to Montreal, 1884. Letters by Clara Lady Rayleigh* (London: Whitehead, Morris & Lowe 1885), 48, 51.

46 "The British Association," *Star*, 7 May 1884.

47 Lefroy, "British Association," 5–6.

48 Howarth, *British Association*, discusses, however, the scientific results (122–5), which included the establishment of committees to deal with issues involving West Coast Indians, tidal observations, and fisheries.

49 Lefroy, "British Association," 3.

50 "Making Ready," *Daily Witness*, [early August] 1884; "The British Association," *Gazette*, [late July] 1884; "The British Association," *Herald*, [late July] 1884; *Quebec Chronicle*, [August] 1884 (MBW: Large scrapbook, 70). See Jack Morrell and Arnold Thackray, *Gentlemen of Science: Early Years of*

the *British Association for the Advancement of Science* (Oxford: Clarendon Press 1981), on the New Museum and the Oxford meeting, 394–5, 407.
51 *Witness,* 28 Aug. 1884.
52 *Herald,* 29 Aug. 1884.
53 *Herald,* 29 Aug. 1884.
54 "The British Association," *Gazette,* 4 Sept. 1884.
55 "The British Association," *Gazette,* 6 and 16 Sept. 1884.
56 "The British Association," *Manchester Weekly Times,* 6 Sept. 1884. Rankine Dawson, ed., *Fifty Years of Work in Canada, Scientific and Educational* (Edinburgh and London: Ballantyne, Hanson & Co. 1901), 225. Howarth, *British Association,* 121. "The British Association: Success of the Montreal meeting," clipping (MC: Dawson papers, box 37). *Witness,* 5 Sept. 1884.
57 MBW: Minute Book, BAAS Executive Committee, 18 April 1885, 57–62 (printed elsewhere as an official return).
58 See Lewis Pyenson, "The Incomplete Transmission of a European Image: Physics at Greater Buenos Aires and Montreal, 1890–1920," *Proceedings of the American Philosophical Society* 122 (April 1978): 92–114 (103).
59 MC: JWD to G.M. Dawson, 5 March 1885. In Dawson, *Fifty Years,* Dawson calls the BAAS presidency the greatest honour of his life (229).

CHAPTER 14

1 JWD to Lorne, 1881.
2 Vittorio M.G. de Vecchi, "The Dawning of a National Scientific Community in Canada, 1878–1896," *Scientia Canadensis* 26 (June 1984): 32–58 (on 33).
3 Rankine Dawson, ed., *Fifty Years of Work in Canada, Scientific and Educational* (London and Edinburgh: Ballantyne, Hanson & Co. 1901), 179.
4 JWD to Lorne, 12 Dec. 1881; Lorne to JWD, 29 Dec. 1881.
5 C.D. Day to JWD, 31 Dec. 1881.
6 JWD to Lorne, 2 Jan. 1882.
7 J.G. Bourinot to JWD, 5 Jan. 1882.
8 Daniel Wilson to JWD, 8 Dec. 1881. Some of the same material from the Dawson correspondence appears in a paper written by two of my former research assistants (see Robert Daley and Paul Dufour, "Creating a 'Northern Minerva': John William Dawson and the Royal Society of Canada," *HSTC Bulletin,* 5 [1981]: 3–13). Another treatment of the reasons for the foundation of the Royal Society appears in Peter J. Bowler, "The Early Development of Scientific Societies in Canada," in A. Oleson and S.C. Brown, eds., *The Pursuit of Knowledge in the Early American Republic* (Baltimore: Johns Hopkins University Press 1976), 326–39 (on 333–5).
9 Daniel Wilson to JWD, 6 Jan. 1882.

10 Wilson to JWD, 12 Jan. 1882.
11 Wilson to JWD, 27 May 1882. The reference to Dawson here presumably meant Rev. Aeneas McDonell Dawson of Ottawa, another member of the English literature section.
12 Wilson to JWD, 4 Feb. 1882.
13 JWD to Lorne, 25 Jan. 1882; Lorne to JWD, 19 Jan. 1882; J.G. Bourinot to JWD, 4 Feb. 1882.
14 Bourinot to JWD, 21 March 1882; Wilson to JWD, 20 April 1882.
15 See Daley and Dufour, "Northern Minerva," 7.
16 *The Royal Society of Canada. Inaugural Meeting* (25–7 May 1882), 6–7.
17 Royal Society. *Inaugural Meeting*, 6–7.
18 Royal Society. *Inaugural Meeting*, 6–7.
19 "The Royal Society of Canada," *Gazette*, 29 May 1882.
20 MC: Dawson papers, box 36, folder 22, H.M. Ami, "Addenda to 1895" (an addendum to his biographical sketch and bibliography previously published in the *American Geologist* 26 [1900]: 1–57).
21 CRS 4 (1891): 326.
22 W. Stewart MacNutt, *Days of Lorne* (Fredericton, N.B.: Brunswick Press 1955), 139. Wilson to JWD, 6 June 1882; Lorne to JWD, 4 June 1882.
23 J.M. Lemoine to JWD, 10 June 1882.
24 Bourinot to JWD, 4 Oct. 1882. This was entitled *The Secretary of the Royal Society – A Literary Fraud* (Ottawa: 1882).
25 Bourinot to JWD, 22 Sept. 1882.
26 CRS 1 (1885): 6.
27 Bourinot to JWD, 31 March 1883.
28 JWD to Alexander Murray, 12 April 1882.
29 T.S. Hunt to JWD, 10 Feb. 1885.
30 JWD to Hunt, Feb. 1885.
31 Accordingly, he opposed the election of R.W. Heneker, despite his exemplary contribution to the cause of Protestant education in Quebec (R.W. Hall to JWD, 4 June 1887; JWD to Hall, 7 June 1887).
32 Royal Society, *Inaugural Meeting*, 10–11.
33 CRS 1 (1885): 7.
34 Wilson to JWD, 20 April 1883; Lorne to JWD, 18 April 1883.
35 Lorne to JWD, 26 May 1883.
36 Bourinot to JWD, 29 May 1883.
37 CRS 4 (1891): 328–9.
38 MC: Dawson papers, box 36, folder 22, "Royal Society of Canada." Vittorio de Vecchi understands this invitation in terms of an abortive attempt to democratize the Royal Society of Canada along the lines of the BAAS; see his "Dawning," (on 38).
39 F.W. Dewinton to JWD, 30 May 1882.
40 Dawson, *Fifty Years*, 201.

41 JWD to P-J O. Chauveau, 1 June 1883.
42 "The Royal Society of Canada," *Gazette,* 29 May 1882.
43 Margaret to JWD, 26 April 1886.
44 JWD to G.M. Dawson, 16 Sept., 1 Oct. 1885. Also see Michael Bliss, *Plague: A Story of Smallpox in Montreal* (Toronto: HarperCollins 1991).
45 MC: Dawson papers, box 22, folder 18, Dawson, Autobiography.
46 NA: JWD to John A. Macdonald, 5 March 1885.
47 Dawson, "Notes on the Meeting of the British Association in Birmingham, 1865," *CN* ns 2 (1865): 9.
48 JWD to G.M. Dawson, 5 March 1885.
49 For example, see Lord Landsdowne to JWD, 25 June 1886.
50 JWD to James Ferrier, 1 Nov. 1883.
51 JWD to Charles Tupper, 17 Aug. 1885; JWD to G.M. Dawson, 19 Aug. 1885.
52 NYL: JWD to James Hall, 8 May 1886; Louisa Carpenter to JWD, 17 July 1886. Yet Dawson returned to England ten years later for the marriage of his son Rankine; at that time, he also attended the British Association meeting in Liverpool (Dawson, *Fifty Years,* 283.)
53 JWD to Arthur Atchison, 15 March 1886; T. Martineau to JWD, 1 June 1886.
54 See MC: Dawson papers, box 37, esp. Peter Bell to JWD, [April?] 1858, for discussion and primitive sketches of Dawson coat of arms. (The resemblance between the Dawson and McGill crests is striking.)
55 M.M. Dawson to JWD, 1886.
56 See obituaries in *Telegram* (Philadelphia), 20 November 1899; *Journal* (Boston), 22 Nov. 1899. An entire scrapbook on the British Association meeting is kept in MC: Dawson papers, box 77. The newspaper clippings referenced below are taken from this scrapbook as well. *Birmingham Gazette,* 2 Sept. 1886.
57 *Hastings Advertiser,* 16 Sept. 1886.
58 Lorne to JWD, 17 Sept. 1886.
59 *Birmingham Gazette,* 2 Sept. 1886.
60 *Journal of Gas Lighting, Water Supply, and Sanitary Improvement,* 7 Sept. 1886.
61 "Address by Sir J. William Dawson," *Report of the Fifty-sixth Meeting of the British Association for the Advancement of Science; held at Birmingham in September 1886* (London: John Murray 1887), 5.
62 *Manchester Guardian,* 2 Sept. 1886.
63 Edward Hull to JWD, 9 Sept. 1886.
64 T.H. Huxley to JWD, 27 Sept. 1886.
65 *Figaro* [English edition], 11 Sept. 1886.
66 Thomas Martineau to JWD, 8 Oct. 1886.
67 Hodder & Stoughton to JWD, 2 Sept. 1886; C.A. Swanston to JWD, 5 Sept. 1886.
68 Archie Duff to JWD, 8 Sept. 1886.

69 JWD to John A. Macdonald, 24 Sept. 1886; JWD to Charles Tupper, 24 Sept. 1886.
70 Herman L. Fairchild, *The Geological Society of America, 1888–1930* (New York: Geological Society of America 1932), 59.
71 Edwin Butt Eckel, *The GSA: Life History of a Learned Society* (GSA Memoir 155: 1982), 7.
72 Eckel, GSA, 8–10.
73 H.S. Williams to JWD, 10 June 1882.
74 J.J. Stevenson to JWD, 22 March 1889.
75 G.M. Dawson to JWD, 31 March 1889.
76 "History of the International Geological Congress," *Geotimes* 5 (1960): 32; T.S. Hunt to JWD, 4 Nov. 1876.
77 William E. Eagan, "The Debate over the Canadian Shield, 1880–1905," *Isis* 80 (1989): 232–53 (on 240).
78 *Geological Magazine*, ns Decade 3, 1 (1884): 432. "History," *Geotimes*, lists the meeting venues.
79 See reports of the proceedings of the congresses in the *Geological Magazine*; for example, in ns Decade 2, 6 (1879): 186–9; ns Decade 2, 8 (1881): 551–60.
80 "History," *Geotimes*; William Thurston, "The First Geological Congress," *Geotimes* 13 (1968): 16. Apropos international scientific congresses, see Elisabeth Crawford, *Nationalism and Internationalism in Science, 1880–1939* (New York: Cambridge University Press 1992), 38–41.
81 "Exhibition," *Encyclopedia Britannica*, 11th ed. (1911).
82 Thomas Weston to JWD, 6 Oct. 1877.
83 James Hall to JWD, 21 June, 15 July 1881.
84 JWD to Frazer, 3 June 1886, 27 Jan. 1887.
85 Hall to JWD, 20 Sept. 1887.
86 Eugène Renevier to JWD, 7 May 1887.
87 [JWD], *CN* 2 (1857): 356.
88 JWD, "Imperial Geological Union," *Nature*, 16 June 1887, 146–7; RS: JWD to Stokes, 17 Feb. 1887.
89 RS: JWD to Stokes, 17 Feb. 1887.
90 MC: Dawson papers, box 12, folder 10, press clipping from *Dundee Advertiser* and *Manchester Examiner*.
91 Selwyn to JWD, 12 March 1887.
92 Lorne to JWD, 29 March 1887.
93 F.B. Wyngdon to JWD, 28 June 1887. This was a preliminary meeting leading up to the creation of the Australasian Association for the Advancement of Science (ultimately, ANZAAS). See Roy MacLeod, ed., *The Commonwealth of Science: ANZAAS and the Scientific Enterprise in Australasia, 1888–1988* (Melbourne: Oxford University Press 1988), 33.
94 MC: Box 1904, folder 10.

266   Notes to pages 203–9

95 T.M. Hughes to JWD, 7 April 1887; Edward Hull to JWD, 20 March 1887; Stokes to JWD, 27 May 1887.
96 RS: JWD to Stokes, 9 June 1887.
97 T.M. Hughes to JWD, 23 Sept. 1887; M. Foster to JWD, 1 Nov. 1887, Dec. 1887.
98 Crawford, *Nationalism*, 38.

CHAPTER 15

1 Epitaph from Dawson's tombstone in Mount Royal Cemetery; (Rev. 14: 13).
2 *Gazette*, 22 Nov. 1899.
3 *Journal* (Boston), 20 Nov. 1899; *Times* (London), 20 Nov. 1899.
4 *The Outlook* (London), 25 Nov. 1899.
5 Rankine Dawson, ed., *Fifty Years of Work in Canada, Scientific and Educational* (London and Edinburgh: Ballantyne, Hanson & Co. 1901), 294.
6 See Stanley Frost, *McGill University for the Advancement of Learning, 1: 1801–95* (Montreal: McGill-Queen's University Press 1980), on all these developments.
7 Hugh MacLennan, ed., *McGill: The Story of a University* (London: Allen and Unwin 1960), 8–9.
8 B. de La Brière to Margaret Dawson, 11 Dec. 1899.
9 *Daily Sun* (St. John, N.B.), 20 Nov. 1899.
10 MRB: Dawson pamphlet collection, "Inaugural discourse," November 1855, p. 4.
11 *The Outlook* (London), 25 Nov. 1899.
12 MC: Dawson papers, box 22, folder 18, Autobiography.
13 JWD to James Hall, 8 May 1886.
14 JWD to Hall, 8 May 1886.
15 Dawson, Autobiography.
16 Clifford Holland, "First Canadian Critics of Darwin," *Queen's Quarterly* 88 (1981): 100–6, (on 105).
17 Anna D. Harrington to G.M. Dawson, [Nov. 1900].
18 Anna to George, 12 Feb. 1901.
19 Rankine to George, 30 Nov. 1900; Margaret to George, 26 Nov. 1900.
20 Anna to George, 18 Feb. 1901.
21 Rankine to George, 1 Dec. 1899.
22 Anna to George, 17 Dec. 1900.
23 Anna to George, 21 Nov. 1900.
24 Margaret to George, 11 Jan. 1900.
25 Anna to George, 11 June 1900.
26 George to Anna, 20 Jan., 12 Feb. 1901.
27 George to Anna, 27 Sept. 1899, 16 Dec. 1900, 9 Feb. 1901.

28 Anna to George, 11 June, [Nov.], 21 Nov., 17 Dec. 1900.
29 Eva to George, 31 Jan. 1901.
30 George to Anna, 18 Jan., 23 Feb. 1901. Anna to George, 22 Jan. 1901.
31 Anna to George, 28 Nov. 1900, 20 Feb. 1901; George to Anna, 23 Feb. 1901.
32 Margaret to George, 22 Feb., 4 Nov. 1900.
33 Anna to George, 6 Nov. 1900.
34 Rankine to George, 12 Jan. 1900; Anna to George, 21 June 1900.
35 Anna to George, 21 June, 21 Nov. 1900; George to Anna, 14 Aug. 1900.
36 Rankine to George, 12 Jan. 1900.
37 Anna to George, 21 June 1900, 25 Feb. 1901.
38 George to Anna, 13 Feb. 1901.
39 Anna to George, 11, 21 June; 19 Sept.; 6, 21 Nov. 1900.
40 George to Anna, 12 Feb. 1901.
41 Anna to George, 22 Jan 1892.
42 Eva to George, 31 Jan. 1901. Anna to George, 18 Feb. 1901.
43 Anna to George, 25 Feb. 1901.
44 Margaret to George, 10 Oct., 14 Dec. 1900.
45 Anna to George, 6 Nov. 1900.
46 George to Anna, 13 Feb. 1901.
47 Anna to George, 21 Nov. 1900.
48 Dawson, Autobiography.

Index

Abbott, John Joseph, 58
Academy of Natural Sciences (Philadelphia), 194
Acadian Geology, 104, 139, publication of, 33–7, reception of, 36–7; (2nd ed.), 52, 85 152–7, reviews of, 156–7; (3rd ed.), 139, 157–8; (4th ed.), 158–9
Acland, Henry, 125
Agassiz, Alexander, 181
Agassiz, Louis, and glaciers, 138, 153; and Edinburgh chair, 41, 43, 47, 48–9; mentioned: 130, 132, 139, 182
Air-breathers of the Coal Period, 151
Airy, G.B., 109
Albert, Prince, 175
Alison, William P., 50–1
Allen, Hugh, 61, 163
Allman, George: and Edinburgh chair, 47, 48, 49–50, 51, 52
American Association for the Advancement of Science (AAAS): and geology, 200; Montreal meeting (1882), 169–71, 178–9, 180–3

American Geologist, 200
Anderson, William James, 45
Andrews, Edmund, 139
Archaia, 82, 83, 121–5
Archibald, Charles D., 33
Argyll, Duke of, 45, 50, 80
Armstrong, George Frederick, 63
autobiography of JWD. *See Fifty Years*

Bache, Alexander Dallas, 170
Baconionism, 130–1
Bailey, Loring Woart, 154, 155, 156, 173
Bailly, Jules F.D., 71
Bain, Francis, 173
Baird, George, 84
Baird, Spencer, 56, 68
Bakerian lecture, JWD's 1870, 101–2, 108–18
Balfour, John Hutton, 23
Ball, Sir Robert, 187
Barnston, George, 177
Barnston, James, 60, 167
Barrande, Joachim, 129
Bayne, James, 45
Baynes, William, 54
Beaudry, Jean-Louis, 78
Bell, Peter, 84; and Edin-

burgh chair, 48–9, 50, 51
Bell, Robert, 96
Bigsby, John Jeremiah: as anti-Darwinist, 103, 111, 126, 134; JWD assists on his *Thesaurus* 102–3; as JWD's friend and advisor, 8, 86, 94, 100–3, 110, 111, 116, 117, 128; mentioned, 115, 138
Billings, Elkanah, 79, 104, 151, 171, 177
biography, writing of, 9, 10, 11
"Birkenshaw." *See* Little Métis
Blackman, Charles, 61
Blackwell, E.T., 60
Bourinot, John George, 191, 192, 195, 196
Bourinot, Marshall, 156
Bovey, Henry, 59, 63
Brewster, David, 80, 83
British Association for the Advancement of Science (BAAS), 24, 202; Glasgow meeting (1855), 52, 53; Montreal (1884), 183–9; Birmingham (1886), 197–200
Brown, Richard, 25, 32, 33, 45, 160, 161

Bunbury, Charles, 25
Burckhardt, Jakob, 120
Butler, Samuel, 165–6
Brush, George, 181, 183

Cambridge University, 200
Canadian Ice Age, 140
Canadian Institute, 174, 191
Canadian Journal of Industry, Science, and Art, 37
Canadian Naturalist, 171–3, 176, 193–4
Canadian Record of Science, 172
Carpenter, Philip H., 147
Carpenter, Philip P., 65, 177
Carpenter, William B., 43, 147, 183, 199; and *Eozoön Canadense*, 140, 141, 143, 145, 146, 147
Carruthers, William, 86, 110, 115, 118
Carter, Henry James, 146
Caswell, Alexis, 170
Chalmers, Robert, 159
Chambers, Richard, 46
Chapman, Edward J., 37, 124, 170
Charlesworth, Edward, 71
Chauveau, Pierre-Joseph-Oliver, 192, 196, 197
Cherriman, John Bradford, 194
Christison, Robert, 81, 83, 84
Collingwood, Cuthbert, 127
Colonial Exhibition, 198
Confederation (Canada), JWD on, 73, 75
Council of Public Instruction (Quebec), 74, 76–7
Cramp, Thomas, 185
Craven, E.J., 87, 126–7
Credner, K. Hermann, 141
Crosby, William O., 147
Curry, Thomas, 68, 70–1

Dallas, William Sweetland, 146
Damon, Robert, 141
Dana, James Dwight, 121, 129, 137, 170, 182; and *Eozoön Canadense*, 140, 146, 147
Darwin, Sir Charles, 38, 48, 114, 125–7, 140
Davin, Nicholas Flood, 195
Davis, W.H.A., 167
Davy, Humphry, 109
Dawn of Life, 146, 201
Dawson, Anna (daughter), 11, 65, 92, 92, 93, 95–6, 207; and *Fifty Years*, 207, 208–11
Dawson, Eva (daughter), 92–3; and *Fifty Years*, 208, 209, 210–11
Dawson, George Mercer (son), 11, 63, 87, 92, 96, 139, 173, 201; assists JWD, 111, 162, 181, 183; career of, 93–5; and *Eozoön Canadense*, 147; and *Fifty Years*, 208–10; and Redpath Museum, 67–8
Dawson, James (father), 4, 16–17, 18, 31, 37, 54, 55
Dawson, James Cosmo (son), 93
Dawson, Sir John William
– early years, 4, 16–17, 18; early education, 19, 20–21, 27, 28; Pictou natural history, 22–3; science interests of parents, 22
– as educator: and Confederation (Canada), 73, 75; lecturer, 28–9, 59, 60, 87, 199; philosophy of, 3, 56, 59, 73–4; summarized, 205–6
– life events: birth, 4; studies at University of Edinburgh, 23, courts and marries Margaret Mercer, 26–8; works as Nova Scotia's superintendant of education, 29–31; appointed to Ryerson Royal Commission on Education, 31; hired to do geological survey of Cape Breton, 33; applies for Edinburgh chair, 44–52; fellow of Geological Society, 52; offered principalship of McGill, 53; AAAS meeting in Montreal, 170–1; fellow of Royal Society (London), 108; applies for Edinburgh principalship, 79–85; Bakerian lecture, 110, Bakerian manuscript rejected, 112–17; turns down geology chair at Princeton, 89; creates Royal Society of Canada, 190–4; organizes AAAS meeting in Montreal, 180–3; Redpath Museum opens, 68–72, 182; travels to Europe and Middle East, 184; BAAS meeting in Montreal, 183–9; knighted, 189, 198; president of BAAS meeting in Birmingham, 189, 197–200; death, 204, 207
– as parent, 91–9
– personality, 6–11, 29–30, 76–7, 91–2, 136–7, 162, 197; Hugh McLennan on, 205; others on, 81, 82, 84; Rankine Dawson on, 209–10; youth, 15, 16, 18, 25, 27, 120
– as popularizer of science, 86, 108, 111, 118, 127–9, 168–9; books, 36, 145, 152–3; fosters social harmony, 175–6; personal reasons for, 118, 119; and scientific credibility, 119, 120, 125, 129
– religion and science: anti-Darwin, 106, 117, 119, 125–6, 127, 129–35; bible scholar, 27, 28, 83, 121–3
– as scientist, 6, 117–18, 149–51, 206–7; Baconian, 137, 168; *Eozoön*

Index

Canadense 128, 137, 140–8, 173, 201; glacial geology, 137, 138–40, 147; on scientific specialization, 138; uniformitarian, 140, 157
Dawson, Margaret Mercer (wife; née Mercer), 26, 28, 29, 54, 93, 135, 181, 197, 199; and *Fifty Years*, 208, 209, 211
Dawson, Mary (mother, née Rankine), 17, 33, 54
Dawson, Rankine (son), 91, 98–9, 181; and *Fifty Years*, 207–11
Dawson, William Bell (son), 93, 147, 181; career of, 96–8; and *Fifty Years*, 208, 210
Dawson Brothers (publisher), 123, 145, 152, 158, 171–2
Day, Charles Dewey, 59, 72, 76, 78, 191
De La Beche, Henry, 23, 48
De La Brière, Boucher, 205
Denderpeton acadianum, 32
Derby, Orville, 147
De Sola, Abraham, 76, 127
Dobson, George H., 161
Dohrn, Anton, 71
Draper, John William, 127
Drew Theological Seminary, 128
drift ice theory. *See* glacial geology
Drummond, George, 57
Duff, Archibald, 63
Dufferin, Earl of, 78
Duncan, Martin, 111, 114, 118
Dunkin, Christopher, 62

Edel, Leon, 9, 10
Edinburgh, University of: geology chair competition, 85; JWD studies at 4; natural history at, 38, 39; principalship competition, 79–85; Regius chair of natural history

competition, 40–52
Egerton, Sir Philip, 36
Eliot, Charles William, 7, 204
Ells, Robert Wheelock, 178
Emmons, Ebenezer, 170
Eozoön Canadense, 128, 137, 140–8, 173, 201

Faraday, Michael, 109, 129
Ferrier, James, 84
Fifty Years of Work in Canada, 10, 12; family feud over, 207–11
Figuier, Louis, 129
Fiske, John, 127
Fleming, John, 41, 42, 50
floating ice theory. *See* glacial geology
Foord, Arthur Humphreys, 178
Foote, A.E., 183
Forbes, Edward, 42, 45, 48, 138
Forbes, James D., 41, 42, 46, 109
Fraser, Donald, 81–2, 193
Fraser, Persifor, 201
Fritsch, Anton, 144
Frost, Stanley, 84
Frothingham, George H., 62–3
Fry, Roger, 11

Geikie, Archibald, 80, 85
General Mining Association, 32–3, 160
Geological Society of America, 200–1
Geological Society of France, 207
Geological Society of London, 24, 32, 103, 126
Geological Survey of Canada, 104, 138–9, 140, 160, 172, 198; and *Acadian Geology*, 157, 158; moves to Ottawa, 63, 66, 68, 77–9, 178
Geology: and *Canadian Naturalist*, 172–3; glacial, 137, 138–40, 147; JWD's

efforts to internationalize, 200–3. *See also Acadian Geology*
Geology of Nova Scotia, New Brunswick, and Prince Edward Island or Acadian Geology. See Acadian Geology (4th ed.)
Gibb, Charles, 64, 175
Gilman, Daniel Coit, 7, 204
Girdwood, Gilbert Prout, 63
glacial geology, 137, 138–40, 147
Gould, Augustus, 182
Grant, Alexander, 81, 83, 84–5
Grant, Anna, 164
Gray, Asa, 64, 117, 133–4, 138, 169, 181, 182, 187
Green, William Henry, 87, 88
Greenshield, David, 58
Greenshields, Edward, 58
Greenshields, J.B., 81
Gregory, William, 23, 46
Grey, Sir George, 47
Gümbel, C.W., 141, 144
Guyot, Arnold Henry, 89, 170

Hahn, Otto, 146
Haliburton, Robert, 161
Hall, James: friendship with JWD, 100, 103–5; and *Paleontology of New York*, 104–5; mentioned, 68, 129, 140, 151, 170, 182, 200, 202
Hamel, Père, 195
Hamilton, Mark, 62
Hankins, Thomas, 11
Harrington, Bernard James, 63, 65, 92, 95–6, 181; and Redpath Museum, 68, 70
Harrington, Clara, 11
Haughton, Samuel, 141
Hay, George, 64
Head, Edmund, 31, 45, 72, 84–5
Henry, Joseph, 170, 182

272   Index

Herdman, Andrew W., 45
Herschel, J.W., 109
Hicks, F.W., 181
Hicks, Henry, 183
Hill, Albert J., 151
Hingston, W.H., 180
Hinks, William, 105
historiography of JWD, 5, 8, 9, 135, 207
history of science: JWD on, 206
Hitchcock, C.H., 200
Hitchcock, Edward, 122, 170
Hodge, Charles, 88–9
Holmes, Andrew Fernando, 65
Home, David Milne, 80, 81, 139
Honeker, Richard W., 77
Honeyman, David, 198
Hooker, Sir Joseph Dalton, 110, 111, 113–14, 117, 118, 130
Hooker, William J., 45
Horner, Leonard, 45, 50–1
Hovey, Rev. H.C., 182
Howe, Henry Aspinall, 60
Howe, Joseph, 45, 164
Howell, Edwin, 68–9
Hughes, Thomas McKenney, 203
Hull, Edward, 200, 203
Hunt, Thomas Sterry, 182; and *Eozoön*, 141, 144, 145; mentioned, 90, 95, 170, 180, 182, 183, 185, 186, 192, 195
Huxley, Thomas Henry: and Edinburgh chair, 42, 43, 47, 52; and *Eozoön*, 140, 146; mentioned, 105, 111, 126, 127, 130, 133, 150, 199, 200, 202
Hyatt, Alpheus, 68, 132–3
Hylerpeton Dawsoni, 150

International Congress of Geologists, 200–2

Jameson, Laurence, 42
Jameson, Robert, 23, 38–9,
56; and Edinburgh chair 42, 44
Jardine, Sir William, 42
Joggins, the (Nova Scotia), 23, 24, 25, 32
Johnson, Alexander, 60, 181
Johnston, George, 46
Jones, Rupert, 144, 183
Judd, J.W., 183
Jukes, Joseph Beete, 50

Kelland, Phillip, 170
Kemp, Rev. Alexander, 124, 128–9
Kew Gardens, 65
King, William, 140, 141, 142, 143, 144, 145, 146, 147
Kingston, George Templeman, 60–1
Kohut, Thomas, 10
Krantz, August, 68
Kuetzing, Paul, 68

Lacoe, R.D., 68
Lansdowne, Lady, 187
LeConte, Joseph, 87, 170
Leisure Hour, 103, 128, 129
Le Marchand, Sir Gaspard, 45
Lemoine, James M., 72
Lesley, Peter, 170
Lesquereux, Leo, 182
Lewis, Henry Carvill, 139
Libbey, Jonas Marsh, 87, 128
Little Métis (Quebec), 7, 55, 96, 149
Lizars, William Home, 34
Lockyer, Norman, 109
Logan, William, 23–3, 33, 45, 64, 67, 145, 150; and *Eozoön*, 140, 145; mentioned, 4, 79, 80, 101, 110, 169, 170, 177
Lonsdale, William, 25
Lorne, Marquis of, 184, 190–2, 193, 196; mentioned, 139, 195, 203
Lyell, Sir Charles, 24, 25, 32, 120, 132, 138, 145,
199; and *Acadian Geology*, 33, 155, 156; Edinburgh chair competition, 43–4, 45–6, 46–50, 51, 52; Edinburgh principalship, 80, 82; JWD's Bakerian lecture, 110, 113, 116; other assistance to JWD, 33, 75, 85, 100, 141, 150; mentioned, 4, 8, 23, 108, 109, 115, 117, 122, 130, 138
Lyman, Henry H., 180
Lyon, Victor, 172

MacAuley, James, 129
McCord, David, 174
McCosh, James, 86, 87, 88, 89
McCulloch, Thomas, 18–19, 22, 23, 37, 119, 127
Macdonald, John A., 61, 77, 78
McGill University, 54–60; agricultural sciences, 61–2; applied sciences, 61–3, 78; BAAS meeting at, 187–8; Burnside Hall, 56; JWD's major accomplishments, 5, 72, 204–5; Macdonald Building, 188–9; natural history, 63–72; Normal School, 74; Peter Redpath Library, 189; Peter Redpath Museum, 65–72, 182, 187; physical sciences, 60–1; Workman Building, 189
McLennan, Hugh, 185
McLeod, Clement H., 60, 61
Macoun, John, 64
Martineau, Thomas, 199, 200
Masson, David, 81
Masson, L-F-R, 79
Matthew, George Frederick, 154, 158, 173
Maxwell, James Clerk, 109
Meek, Fielding, 103
Mercer, Margaret. *See* Dawson, Margaret

meteorology, 60
Micmac, 153
Miles, Henry Hopper, 76
Miller, Hugh, 43, 50, 82, 122, 129
Miller, Samuel A., 139
Milne-Edwards, Henri, 48
Mining: 94, 95, 160–4; General Mining Association, 32–33; McGill, 61– 3
Möbius, Karl, 146, 147
Molson, Anne, 66
Molson, John Henry, 62–3, 67, 70, 71, 181, 187
Molson, Louisa, 70
Molson, William, 65
Molson family, 57–8, 163, 183
Montreal: described, 16, 53–4, 165–6, 186
Moore, Charles, 71
Morgan, H.J., 195
Moseley, Henry N., 188
Murchison, Sir Roderick, 48, 80, 82, 108, 154, 155
Murray, Alexander, 195
Museums: Natural History Society of Montreal, 167–9, 175, 177–8, 179; Peter Redpath Museum, 65–72, 182, 187

Natural History Society of Montreal, 60, 166–9, 171–9, 188, 197
Newberry, John Strong, 139
Newcomb, Simon, 187
Nicholson, Henry Alleyne, 147–8, 173
Nicol, James, 34, 49
Notes on the Post-Pliocene Geology of Canada, 139

O'Brien, Charles, 8, 147
Oliver and Boyd (publisher), 34, 35, 158
Origin of Species, 129–34, 151–2, 155
Origin of the World, 87, 123, 125–7
Owen, Richard, 32, 47, 86, 109, 150

Owens College, 59, 62

Paleobotany, 107–8, 112–13, 116–17, 118
Paley, William, 8, 21
Palmerston, Lord, 47
Papineau, Joseph, 175
Pattison, J.R., 128, 199
Penhallow, David, 64, 65, 70, 95, 205
Peter Redpath Museum, 65–72, 182, 187
Phillips, John, 23, 36, 41, 47
Philosophical Transactions, 108–9, 113, 116
Pictou (Nova Scotia), 4, 15, 16, 27; JWD lectures at, 23, 25; iron mining, 161–3
Pictou Academy, 18–22, 66
Pierce, Benjamin, 170
Pinnock, William, 120
Plant, John, 141
Playfair, Lyon, 81
Poole, Henry S., 33, 45
Prentice, Edward A., 161–3
Prestwich, Joseph, 139
Primrose, Howard, 163
Princeton College, 86–9
Princeton Review, 87, 128
Princeton Theological Seminary, 86–7, 88
Pupa vetusta, 32

Quatrefages, Armand, 48

Rae, John, 170, 184
Ramsay, Andrew Crombie, 108, 144, 153, 170
Rankine, Mary. *See* Dawson, Mary
Rathbun, Richard, 68, 147
Rayleigh, Clara, 186
reciprocity treaty (Can.-U.S.), 161
Redpath, Mrs John, 187
Redpath, Peter, 57–8, 62–3, 163; and museum, 67, 68, 70, 72, 182
Redpath family, 199
Redpath Museum, 65–72, 182, 187

Relics of Primeval Life, 148
Religious Tract Society, 103, 128
Rennie, William, 167
Roberts, George, 71
Robertson, Andrew, 187–8
Rogers, William B., 182
Ross, W.J., 33, 164
Rowney, Thomas H., 141, 142, 143, 145, 146, 147
Royal College of Chemistry, 62
Royal Institution, 111
Royal Society (London), 78, 103, 111, 114–15, 151, 194–6, 202–3, 207; JWD's 1870 Bakerian lecture, 101–2, 108–18; *Philosophical Transactions*, 108–9, 113, 116
Royal Society of Canada, 192–7
Ryerson Royal Commission on Education, 31

Sabine, Edward, 109, 110
Schurman, Jacob, 87
Schuster, Arthur, 109
Scudder, Samuel, 68, 139, 173, 181
Seaman, Berthold, 170
Selwyn, Alfred Richard Cecil, 78, 79, 96, 177, 193, 203
Sharpey, William, 81
Shore, Miles, 10
Silliman, Benjamin, 170, 182
Simpson, James, 80–1, 83, 84
smallpox in Montreal, 197–8
Smallwood, Charles, 60, 169, 173
Smith, James, 127
Smithsonian Institution, 65
Sorby, Henry Clifton, 146
Sowerby, George B., Sr., 25
Spencer, Herbert, 127, 199
Spencer, Joseph W.W., 173

Stevenson, J.J., 201
Stevenson, Major S.C., 180
Stokes, George, 113, 115, 202
Story of Earth and Man, 123, 128, 145
Sutherland, William, 60
Syme, James, 81

Thomas, Francis W., 180
Thomson, J.J., 109
Thomson, William, 81, 109
Torrance, J.F., 183
Traill, Thomas Stewart, 42
Tupper, Charles, 191
Turner, William, 84
Tyndall, John, 109, 111, 126, 127, 138, 200

uniformitarianism, 138, 140, 157
Union Theological Seminary (New York), 87, 127

University of Toronto, 105–6

Vail, Isaac Newton, 139
Vallance, T.G., 118
Vaux, William S., 182
Verrill, Addison Emery, 139
Vroom, James, 64

Walcott, Charles Doolittle, 134
Wallace, Alfred Russel, 199
Ward, Henry, 68, 71, 172
Weir, George, 76
Wellhausen, Julius, 120–1
Werner, Abraham Gottlob, 39
Wernerian Society (Edinburgh), 23
Weston, Thomas Chesmer, 178
Whewell, William, 121
White, Thomas, 68, 185

White, Walter, 115
Whiteaves, J.F., 167, 173, 176, 177–8
Whitfield, R.P., 68
Williams, Henry Shaler, 134
Williamson, William Crawford, 6–7, 100, 110, 146
Wilson, Daniel, 57, 100, 105, 191–2; mentioned, 170, 176, 181, 185, 193
Winchell, Alexander, 134
Winslow-Sprague, Lois, 9
Wood, H.H., 57
Woodward, Henry, 68
Woolf, Virginia, 11
Workman, Thomas, 63
Workman, William, 58

X-Club, 111
Xylobius sigillariae, 32

Yale College, 62
Young, William, 45